AF567135

Rapid Characterization of Microorganisms by Mass Spectrometry

ACS SYMPOSIUM SERIES **1065**

Rapid Characterization of Microorganisms by Mass Spectrometry

Catherine Fenselau, Editor
University of Maryland

Plamen Demirev, Editor
Johns Hopkins University

Sponsored by the
ACS Division of Analytical Chemistry

American Chemical Society, Washington, DC

Distributed in print by Oxford University Press, Inc.

Library of Congress Cataloging-in-Publication Data

Rapid characterization of microorganisms by mass spectrometry / Catherine Fenselau, editor, Plamen Demirev, editor.
p. cm. -- (ACS symposium series ; 1065)
"Sponsored by the ACS Division of Analytical Chemistry."
Includes bibliographical references and index.
ISBN 978-0-8412-2612-8 (alk. paper)
1. Microorganisms--Identification. 2. Mass spectrometry. 3. Microbiological chemistry. I. Fenselau, Catherine, 1939- II. Demirev, Plamen. III. American Chemical Society. Division of Analytical Chemistry.
QR69.M33R37 2011
571.2'9--dc23

2011017703

The paper used in this publication meets the minimum requirements of American National Standard for Information Sciences—Permanence of Paper for Printed Library Materials, ANSI Z39.48n1984.

Distributed in print by Oxford University Press, Inc.

PRINTED IN THE UNITED STATES OF AMERICA

Foreword

The ACS Symposium Series was first published in 1974 to provide a mechanism for publishing symposia quickly in book form. The purpose of the series is to publish timely, comprehensive books developed from the ACS sponsored symposia based on current scientific research. Occasionally, books are developed from symposia sponsored by other organizations when the topic is of keen interest to the chemistry audience.

Before agreeing to publish a book, the proposed table of contents is reviewed for appropriate and comprehensive coverage and for interest to the audience. Some papers may be excluded to better focus the book; others may be added to provide comprehensiveness. When appropriate, overview or introductory chapters are added. Drafts of chapters are peer-reviewed prior to final acceptance or rejection, and manuscripts are prepared in camera-ready format.

As a rule, only original research papers and original review papers are included in the volumes. Verbatim reproductions of previous published papers are not accepted.

ACS Books Department

Contents

Indexes

Chapter 1

Rapid Characterization of Microorganisms by Mass Spectrometry: An Overview

Catherine Fenselau[1] and Plamen Demirev[*,2]

[1]Department of Chemistry and Biochemistry, University of Maryland, College Park, MD
[2]Applied Physics Laboratory, Johns Hopkins University, Laurel, MD 20723
[*]plamen.demirev@jhuapl.edu

Mass spectrometry approaches for rapid characterization of microorganisms date back more than thirty five years. Recent instrument and methods developments have brought to the fore new and exciting application in a number of fields, reviewed in individual chapters of this book.

This book contains chapters from participants in a recent symposium, organized by the Division of Analytical Chemistry during the ACS National Meeting in Washington DC in August, 2009, as well as several other active researchers in the field. The book covers aspects of mass spectrometry (MS) applications for microorganism characterization in several fields: biodefense, clinical diagnostics, food safety, environmental monitoring, and chemotaxonomy/biosystematics. The diverse list of contributors - from academia, government as well as industry – present multi-faceted and broad perspectives on the subject to be presented.

Through the last thirty-five years MS has continuously provided profiles of chemical ensembles that are characteristic of different species of microorganisms. Such analyses have benefited from the strengths of mass spectrometry—speed, sensitivity, specificity and automation—and have also been influenced by its limitations, including costly MS instrumentation. The progress in mass spectrometry ionization methods and mass analyzers has been reflected in the particular approaches and protocols for microorganism analysis developed during this time. Although phospholipids and low mass metabolites were initially recognized as species specific biomarkers (*1*, *2*), matrix-assisted laser desorption/ionization (MALDI), electrospray and other advances in mass spectrometry have made it possible to profile and identify proteins and oligonucleotides (*3*) from bacteria. Two complementary strategies have evolved for assigning the identities of bacteria based on protein profiles detected in mass spectra. In one approach, careful culturing and controlled MALDI measurements allow the fingerprint comparison of a sample spectrum to a library of target reference spectra, and has been evaluated recently for use in clinical diagnostics (*4*). The other, a proteomic strategy, does not require a library of reference spectra, but uses bioinformatics (*5*, *6*). Detected proteins are identified by relating suites of masses and partial sequences to a library of genomic sequences. Partial protein sequences can be obtained experimentally using standard bottom up proteomic techniques to identify peptides generated in bacterial lysates (*7*, *8*) or from top down MS/MS analysis of intact proteins (*9*, *10*). Recognizing biomarkers based on sequence tags allows reliable analysis of bacteria in mixtures (*11*) and characterization of bacteria that have been genetically engineered (*12*). Advances in bioinformatics have facilitated the identification of proteins in bacteria that lack a sequenced genome (*13*), and the phylogenetic characterization of such bacteria (*14*, *15*).

Several comprehensive reviews of the field (e.g., (*3*, *16*, *17*)) and two books (*18*, *19*) have appeared recently. Here we list a number of issues that in our opinion are likely to influence future developments in the field:

- The first of these is the need for a standard spectral library that is independent of instrument and manufacturer. A mechanism should be provided for independent researchers to contribute spectra to this library.
- Signature validation and annotation will be required for applications in clinical and regulatory areas.
- Continued sequencing of microorganism genomes should be encouraged, along with contribution of annotated genomes to public databases.
- Readily achievable extensions should be explored, including the detection/analysis of drug resistance in bacteria.
- Time for analysis could be shortened considerably by new methods for analysis of mixtures.
- The community should aspire to develop robust, inexpensive, field-deployable systems that provide "end-to-end" capability, that is, systems that collect and prepare the sample, in addition to introducing it into the mass spectrometer and reporting out analysis of the data. Such systems are more likely to be adapted in hospitals and emergency units.

These and other developments will undoubtedly enhance the successful introduction of mass spectrometry in clinical microbiology and infectious disease diagnostics.

In summary, we are pleased to bring together this set of current reports from leading contributors to the field, which we hope will stimulate further advances in both the technology and its applications.

References

1. Anhalt, J. P.; Fenselau, C. Identification of bacteria using mass-spectrometry. *Anal. Chem.* **1975**, *47*, 219–225.
2. *Mass Spectrometry for the Characterization of Microorganisms*; Fenselau, C., Ed.; ACS Symposium Series; American Chemical Society: Washington, DC, 1994; Vol. 541.
3. Demirev, P. A.; Fenselau, C. Mass spectrometry for rapid characterization of microorganisms. *Annu. Rev. Anal. Chem.* **2008**, *1*, 71–93.
4. Cherkaoui, A.; Hibbs, J.; Emonet, S.; Tangomo, M.; Girard, M.; Francois, P.; Schrenzel, J. Comparison of two matrix-assisted laser desorption ionization-time of flight mass spectrometry methods with conventional phenotypic identification for routine identification of bacteria to the species level. *J. Clin. Microbiol.* **2010**, *48*, 1169–1175.
5. Demirev, P. A.; Ho, Y. P.; Ryzhov, V.; Fenselau, C. Microorganism identification by mass spectrometry and protein database searches. *Anal. Chem.* **1999**, *71*, 2732–2738.
6. Demirev, P. A.; Lin, J.; Pineda, F. J.; Fenselau, C. Bioinformatics and mass spectrometry for microorganism identification: Proteome-wide post-translational modifications and database search algorithms for characterization of intact *H.pylori*. *Anal. Chem.* **2001**, *73*, 4566–4573.
7. Warscheid, B.; Jackson, K.; Sutton, C.; Fenselau, C. MALDI analysis of Bacilli in spore mixtures by applying a quadrupole ion trap time of flight tandem mass spectrometer. *Anal. Chem.* **2003**, *75*, 5608–5617.
8. Sundaram, A. K.; Gudlavalleti, S. K.; Oktem, B.; Razumovskaya, J.; Gamage, C. M.; Serino, R. M.; Doroshenko, V. M. Atmospheric pressure MALDI-MS-MS based high throughput automated multiplexed array system for rapid detection and identification of bioagents. *Proceedings 56th Conference of the American Society for Mass Spectrometry*, Denver, CO, June 1–5, 2008.
9. Demirev, P. A.; Feldman, A. B.; Kowalski, P.; Lin, J. S. Top-down proteomics for rapid identification of intact microorganisms. *Anal. Chem.* **2005**, *77*, 7455–7461.
10. Fagerquist, C. K.; Garbus, B. R.; Williams, K. E.; Bates, A. H.; Boyle, S.; Harden, L. A. Web-based software for rapid top-down proteomic identification of protein biomarkers, with implications for bacterial identification. *Appl. Environ. Microbiol.* **2009**, *75*, 4341–4353.
11. Warscheid, B.; Fenselau, C. A targeted proteomics approach to the rapid identification of bacterial cell mixtures by MALDI mass spectrometry. *Proteomics* **2004**, *4*, 2877–2892.
12. Russell, S.; Edwards, N.; Fenselau, C. Detection of plasmid insertion in *E.coli* by MALDI-TOF mass spectrometry. *Anal. Chem.* **2007**, *79*, 5399–5409.
13. Wynne, C.; Fenselau, C.; Demirev, P. A.; Edwards, N. J. Top-down identification of protein biomarkers in bacteria with unsequenced genomes. *Anal. Chem.* **2009**, *81*, 9633–9642.
14. Teramoto, K.; Sato, H.; Sun, L.; Torimura, M.; Tao, H.; Yoshikawa, H.; Hotta, Y.; Hosoda, A.; Tamura, H. Phylogenetic classification of

Pseudomonas putida strains by MALDI-MS using ribosomal subunit proteins as biomarkers. *Anal. Chem.* **2007**, *79*, 8712–8719.
15. Wynne, C.; Edwards, N.; Fenselau, C. Phyloproteomic classification of unsequenced organisms by top-down identification of bacterial proteins using capLC-MS/MS on an Orbitrap. *Proteomics* **2010**, *10*, in press.
16. Demirev, P.; Fenselau, C. Mass spectrometry in biodefense. *J. Mass Spectrom.* **2008**, *43*, 1441–1457.
17. Ho, Y. P.; Reddy, P. M. Identification of pathogens by mass spectrometry. *Clin. Chem.* **2010**, *56*, 525–536.
18. *Identification of Microorganisms by Mass Spectrometry*; Wilkins C. L., Lay, J. O., Eds.; Wiley-Interscience: New York, 2005.
19. *Mass Spectrometry for Microbial Proteomic*; Shah, H. N., Gharbia, S. E., Eds.; Wiley-Interscience: New York, 2010.

Chapter 2

Rapid Sample Preparation for Microorganism Analysis by Mass Spectrometry

Franco Basile*

University of Wyoming, Department of Chemistry, Laramie, WY 82071
***basile@uwyo.edu**

The time required for rapid microorganism analysis by Mass Spectrometry (MS) should take into account not only the time devoted to perform the measurement, but also the time to prepare the biological sample (including growth and isolation) and perform the data analysis. All MS systems ultimately analyze gas phase ions, and thus, the challenge then becomes on how to convert molecules present within the microorganism into gas phase molecules and subsequently ions that can be analyzed by MS. As a result, the microorganism sample preparation step depends on the microorganism(s) being studied, the type of biomolecule targeted, how the sample is either volatilized or desorbed into the gas phase, ionization mode, how ions are sampled or introduced into the MS vacuum system, mass analyzer type and the level of detection selectivity and specificity required. In this Chapter, an overview is presented on different sample preparation steps used in the context of rapid analysis of microorganisms with different MS techniques.

Introduction

Investigators in the 1970's and 80's using mass spectrometry were limited by the choices of available sample introduction and ionization modes, which at the time were mostly based on thermal desorption (*i.e.*, sample vaporization), Electron Ionization (EI) and Fast Atom Bombardment (FAB). Sample vaporization either employed the hot inlet of a gas chromatograph (GC), a direct insertion probe (DIP) or a Pyrolysis (Py) probe. This fact limited MS analysis to relatively low molar mass and thermally stable molecules. For the analysis of microorganisms, the choice then came down to either extract, derivatize and analyze small molar mass molecules *via* GC-EI-MS or directly analyze intact cells by Direct Probe-EI-MS. The Direct Probe-EI-MS, either through slow heating (*1*) or pyrolysis (*2*, *3*), approach became attractive since in a single step and *in situ* (inside the MS vacuum system) intact cells were lysed, biomolecules digested/degraded to small molar mass biomarker molecules, and biomarkers thermally desorbed for subsequent EI-MS analysis. The process was rapid, taking several seconds (Curie-point Py) to less than 5 minutes to complete a heating cycle, making it attractive for rapid near real-time MS analysis. The resulting mass spectra represented fingerprints or patterns for the microorganism being analyzed, and with the aid of pattern recognition techniques, classification to the species and strain level was possible in controlled sample sets. Analyses employing FAB-MS (*4*) and FAB-MS/MS (*5*) allowed the detection of phospholipid profiles from pretreated microorganisms and crude lipid extracts. The soft character of FAB made possible the detection lipid biomarkers with molar masses between 600 and 1300 amu, thus increasing the specificity of the bacteria detection scheme.

With the advent of Electrospray Ionization (ESI) (*6*) and Matrix Assisted Laser/Desoprtion Ionization (MALDI) (*7*), the analysis of thermally labile high molar mass biological molecules became feasible. Many reviews on ESI (*8–10*) and MALDI (*11–15*) have been published detailing fundamental principles for each ionization technique as well as new applications in the field. Both techniques have been used in the analysis of microorganisms; however, it is MALDI-MS that has found greater application in this field due to its inherently easy "whole cell" sample preparation step. Several advancements in the analysis of whole cells by MALDI-MS will be discussed as well as a survey of methodology for successful whole cell analysis by MALDI-MS.

Finally, the application of recently developed Ambient MS techniques like Desorption-Electrospray Ionization (DESI) and Direct Analysis in Real Time (DART) to the analysis of microorganisms will be covered. In ambient MS the sample (and analyte(s) within) is analyzed in its native state and at atmospheric pressure, that is, the measurement precludes any sample preparation step. This has obvious advantages in the context of rapid analysis of microorganisms as the sample preparation step in some instances consumes up to 50-75% of the analysis duty cycle time.

Microorganism Analysis by Ambient-MS

In any rapid assay, the best sample preparation is ultimately no sample preparation step. Two novel atmospheric pressure ionization techniques DESI-MS (*16*) and DART (*17*) were introduced in 2004. These techniques demonstrated that sample preparation can be eliminated or at least incorporated as a step of the overall ionization/desorption process. Several reviews have been written on the subject that illustrate the utility, mechanism of operation and range of applications of DESI, DART and other variants of Ambient MS (*18–22*). Given the potential of no sample preparation imparted by Ambient MS techniques, only few reports are available on their use for the direct analysis of microorganisms. With Ambient

MS techniques either lipids and/or small molar mass metabolites are the primary biomarkers that have been detected by these approaches. In this regard, it is well documented the usefulness of lipid composition for the classification and identification of microorganisms, which is usually based on fingerprint methods of the detected lipid profiles. As such, Ambient MS approaches currently provide an alternative for rapid microbial lipid profiling.

Ambient-MS of Microorganisms

In DESI-MS of intact bacteria, the sample is simply deposited onto a solid substrate (glass, Teflon, etc) and the dried spot analyzed by DESI-MS. A typical analysis of an aqueous bacterial suspension (~10^8 cells/mL) by DESI-MS would consist in depositing a 1-2 μL aliquot of the cell suspension onto the DESI probe surface (~10^5 cells on probe) and allowed to dry. The DESI source is then turned on and mass analysis achieved in the desired mass range. Usually, several spots are analyzed for replicate measurements. The entire process usually takes 5-10 minutes, depending on the number of replicates. The Cooks laboratory first demonstrated the potential of analyzing intact and untreated bacteria with DESI (*23*) by analyzing freshly harvested bacteria from agar plates (*E. coli* and *Pseudomonas aeruginosa*, both Gram negative bacteria). Samples were dried onto a polytetrafluoroethylene (PTFE; a.k.a., Teflon®) target and DESI mass spectra obtained using 50% methanol/water as the spray solution. The resulting mass spectra, in the mass range of 100-1100 amu, showed different and characteristic mass spectral fingerprints that allowed their differentiation.

Two subsequent reports studied the analysis of bacteria by DESI-MS in more detailed. The first detailed study of DESI-MS of bacteria was also conducted by the Cooks group where they analyzed several strains of *E. coli* and *Salmonella typhimurium* (*24*). Bacteria samples were prepared as water suspensions (10^8 cells/mL) from bacteria isolated and washed from nutrient broth. The reproducibility of the method was demonstrated by measuring replicate DESI-mass spectra for *E. coli* harvested from the same and different cultures. These positive ion mode DESI-mass spectra were dominated by protonated and cation-adduct phospholipids (in the 700-900 amu region), and their dimmers (in the 1350-1550 amu region). Even though, for the purpose of this work samples were grown in media broth, samples collected from agar plates can be directly analyzed without any prewash step (*vide infra*). Overall, sample analysis of the bacterial water suspension can be estimated at less than 5 min, making the DESI-MS approach suitable for near-real time microorganism detection.

A second report by the Basile group showed a detailed study of DESI-MS of bacteria in the low molar mass region (50-500 amu) (*25*). This work confirmed the ability of DESI-MS to obtained rapid mass spectral fingerprints from intact cells, and also expanded the study to included a larger set of bacteria, both Gram negative and Gram positive bacteria, for their differentiation (*E. coli*, *S. aureus*, *Enterococcus*, *B. thuringiensis*, *Bordatella bronchiseptica*, *Bacillus subtilis* and *Salmonella enterica*). In this study, bacteria were washed with water to remove any potential contamination from the media and about 10^7 cells were deposited onto the probe. DESI-mass spectra were measured in both positive and negative ionization modes, the latter producing a larger number of low molar mass biomarker signals with characteristic signals for some free fatty acids and low molar mass metabolites. Data were collected over a period of 3 days and the resulting DESI-mass spectral fingerprints compared using pattern recognition (Principal Component Analysis, PCA). The analysis allowed for complete differentiation of all tested bacteria, with the notable exceptions of *B. subtilis* and *S. aureus* which were not differentiated from each other in this study. It was also noted that changes in the growth media preparation protocol caused an increase in the intra-group variability (as measured by PCA); however, the inter-group variability was maintained to provide complete differentiation among the bacteria studied (with the noted exceptions above). Finally, in a subsequent study by the Cooks group (*26*) DESI-MS was employed to obtain high quality bacterial lipid profiles in both positive and negative ionization modes. In this study, bacteria suspensions were prepared in 70% ethanol, rather than water and deposited onto glass slides for DESI-MS analysis. Principal component analysis of the resulting DESI-mass spectra showed clear differentiation of four species of bacteria studied: *S. aureus*, *Bacillus subtilis*, *E. coli* and four strains of *Salmonella.* Differentiation of the bacteria studied *via* their DESI-generated lipid profiles was also possible even when grown in different media. These results are in agreement with previous studies on the effect of growth conditions on lipid bacterial profiling employing Py-MS of intact bacteria (*27–29*). The observed prevalence of phospholipids in the DESI-mass spectra of intact bacteria is most likely due their accessibility by the DESI solvent and their higher surface activity relative to that of proteins and oligonucleotides.

DESI-MS was also applied to the direct analysis of *Bacillus subtilis* grown as biofilms on filter paper (*30*). The bacteria were grown on a filter paper deposited on top of agar media. After the incubation period, the filter paper, now containing the biofilm, was taken directly for DESI-MS analysis, that is, no bacteria suspensions were made prior to analysis. High quality mass spectra were obtained in both the positive and negative ionization modes, with a strong signals corresponding to the cyclic lipopeptide Surfactin(C15). This study clearly demonstrated the capability of DESI to completely bypass the sample preparation step and perform the sampling, desorption and ionization of bacterial biomarkers directly from untreated microorganisms and under ambient conditions.

During the DESI process the bacterial sample is subjected to a high velocity gas, and thus, this raises concerns about the analysis of pathogenic microorganisms with a method that can potentially aerosolize the sample, creating an obvious bio-hazard situation for the operator. However, the use of an enclosed DESI source to reduce exposure of the analyst to potentially hazardous samples (*31*) should address this issue. The advantages of DESI as a field-deployable sampling/ionization technique for the analysis of microorganisms (pathogens) are obvious (*32*, *33*) and future studies and applications of DESI for the analysis of microorganisms are expected to be geared at establishing its ruggedness, and reproducibility. However, because DESI mostly detects lipid and metabolite biomarkers, microorganism classification and identification is limited to a profile-based approach, which hinders the applicability of DESI-MS to the analysis of microbial mixtures. Here, the ability to perform reactive DESI (*34*, *35*) should expand the type of biomarkers accessible through this rapid and sample preparation-free technique.

Ambient-MS of Microorganisms Using Reagents

In some instances, biomarker detection from whole microorganism has not been possible by direct implementation of Ambient MS techniques. In one instance, the targeted detection of the spore biomarker dipicolinic acid (DPA) (*36–39*) in untreated Bacillus and Chlostridium spores, has not been possible with DESI (Basile laboratory, unpublished results). Moreover, it remains to be demonstrated the ability of DART to detect DPA directly from untreated spores or for the analysis of untreated bacteria. Finally, the detection of proteins from untreated microorganisms with DESI-MS has not been reported, most likely for reasons stated earlier. In all these instances, sample preparation is implemented prior to detection with Ambient MS, *albeit*, either as a very simple step prior measurement or in the form of the addition of a derivatization reagent to the sample. These approaches can still be classified as rapid analyses since the resulting sample is analyzed either as a crude mixture or the sample preparation is done *in situ*, that is, during the ionization/detection process, maintaining the overall analysis time under 10 min.

Dipicolinic acid (DPA, *M* 168) is present in Bacillus spores at about 10% by weight, making it an ideal biomarker for the sensitive detection of spores (*40*). DPA, in addition to α/β small acid-soluble spore proteins (SASP), is extremely important in spore resistance, stability and in protecting spore DNA from damage (*41*). Because of its importance as a biomarker in biodefense applications, a wide variety of analytical techniques have been used for the detection of DPA in spores including liquid chromatography (*42*), Py-MS (*36*), Py-GC-Ion Mobility Spectrometry (*43*), Py-MS and FT-IR (*44*), time-resolved fluorescence spectroscopy (*45*, *46*) and surface enhanced Raman scattering (*47*, *48*) among many others.

For the detection of DPA in spores with ambient MS, the Basile group analyzed crude lyophilized *Bacillus subtilis* spores in their media (soy broth) without resorting to a washing step. The analysis strategy was to combine thermal desorption (TD) with *in situ* methylation by the addition of the reagent tetramethylammonium hydroxide (TMAH) (*36*, *49*, *50*). In this strategy, the sample is deposited in a TD tube along the with the TMAH reagent (<10 μL, 0.1 M). Heating the spore sample with TMAH at temperatures between 250-300 ^{o}C releases the DPA biomarker, which is rapidly methylated at both carboxylic acid groups. The resulting volatile dimethylated DPA (2Me-DPA, *M* 195) derivative is detected by a pneumatically-assisted ESI source located just above and perpendicular to the TD tube containing the spore sample (*51*). Using the Atmospheric Pressure (AP)-TD/ESI approach, the DPA biomarker in spores was detected as the protonated molecule ($(M+H)^+$, *m/z* 196) and sodium ion adduct ($(M+Na)^+$, *m/z* 218) in crude lyophilized samples with growth media, completely precluding any cell washing step prior to analysis. A detection limit for the 2Me-DPA biomarker was estimated at 1 ppm (equivalent to 0.01 μg of DPA deposited in the thermal desorption tube), which corresponded to a calculated detection limit of 10^5 spores deposited or 0.1% by weight spore composition in solid samples (assuming a 1 mg sample size). No fatty acid methyl esters (FAMEs) were detected in this study, most likely due to the known reduction in lipid content during the sporulation period in microorganisms. This trend was also observed in previous analysis of *Bacillus anthracis* spores by pyrolysis *in situ* methylation using a field portable quadrupole ion trap MS system. That is, when *B. anthracis* cells in the vegetative stage were analyzed in this manner, FAME profiles were obtained; however, *B. anthracis* spores only produced prominent signals corresponding to the 2Me-DPA biomarker (*29*).

The Fernandez group used DART (*17*) with *in situ* methylation for the rapid analysis of intact bacteria by ambient generation of FAME profiles (*52*). In this approach, the bacterial sample is mixed with the reagent TMAH and a drop of this mixture is suspended in a glass capillary tube. The addition of TMAH to methylate polar bacterial biomarkers was warranted since DART is known to be best suited for the analysis of volatiles and semi-volatiles molecules (*53*, *54*). The bacteria/TMAH drop is then placed in the path of the DART He gas stream and directly in front of the inlet orifice of the MS system. When the bacteria/TMAH mixture comes in contact with the hot He gas from the DART source (gas temperature varied between 150-500 °C), rapid *in situ* methylation takes place and the resulting volatile FAMEs are detected by MS. Analysis was rapid, with less than 10 min total analysis time for bacterial suspensions, providing a rapid platform for the generation of bacterial FAME profiles under ambient conditions. It remains to be demonstrated, however, the reproducibility of the bacterial FAME profiles generated *via* the DART/TMAH approach, a requirement for the differentiation of microorganisms with the aid of pattern recognition data analysis (*55*, *56*).

As mentioned earlier in this section, because of the high surface activity of microbial phospholipids, they are preferentially detected over other biomarkers like proteins. In fact, investigations on the detection of the viral capsid protein of the *E. coli* bacteriophage MS2 by DESI-MS (*57*) led to a protocol that required the separation of the protein fraction from other biomarkers before the measurement. The cell lysate (*E. coli* cells infected with the MS2 bacteriophage) was fractionated with a 100 kDa molar mass cutoff spin column. The collected crude fraction was deposited onto a glass slide and analyzed by DESI-MS using a solvent consisting of 70% formic acid/acetonitrile generating signals corresponding to the multiply charged MS2 capsid protein. Detection of proteins at a higher signal-to-noise ratio (S/N) has been achieved using liquid sample DESI (*58*); however, it remains to be shown the applicability of liquid sample DESI to the analysis of intact bacteria or crude bacteria lysates.

Aerosol/Single Particle Mass Spectrometry for Microorganism Characterization

Another methodology where no sample preparation is required is single particle MS or bioaerosol MS (BAMS) (*59*) for the characterization of microorganisms in aerosols (*60*) directly by MS. This is an important approach as it has obvious applications in biodefense, where a biological weapon is most likely to be disseminated as an aerosol. Single particle MS for the detection of microorganisms is discussed in detail by Mathias Frank's chapter in the current ACS volume.

Sample Preparation for MALDI-MS Profiling of Microorganisms

MALDI-MS is probably the most widely used MS technique for the analysis of microorganisms (*61–63*). This stems in part from the fact that sample preparation in MALDI is inherently simple, and this simplicity is also carried to the analysis of microorganism. In addition, MALDI-MS can generate information rich profiles of mostly (ribosomal) protein biomarkers from microorganisms. In the first reports of whole cell analysis by MALDI-MS (*64–66*), sample preparation roughly involved spotting a bacterial suspension onto the MALDI plate, mix in the matrix solution, drying the mixture and analysis by MALDI-MS in order to obtain a protein profile of the bacteria. With time, investigators became aware of reproducibility issues in the resulting MALDI-mass spectral profiles (*67*), and thus, a lot of effort (and care) has been placed in the sample preparation step for the analysis of microorganisms by MALDI-MS. The analysis of whole microorganisms by MALDI-MS is analogous to the analysis of a complex (and unknown) mixture of biological molecules and as such the resulting mass spectrum is affected by both sample history and sample preparation protocol. A survey of published protocols for MALDI-MS analysis of peptides and proteins reveals that many different, but equally successful sample preparation protocols have been developed. That is, the MALDI process is not constrained by a narrow set of experimental variables, but rather it is viable under a wide range of parameters and this has led to many "recipes" for sample preparation. This inherent ruggedness of the MALDI process is also true for the analysis of microorganisms with MADLI-MS, and is reflected by the wide range of published protocols (*68–73*) designed to yield reproducible and high quality bacteria mass spectral fingerprints. Although investigators in the field are far from having a unified and widely accepted sample preparation protocol for MALDI-MS of microorganisms, the technique is gaining acceptance in the clinical, medical and environmental fields as a viable research and diagnostic tool for microorganism detection (*74*). In addition, commercial MALDI-MS systems with specifically designed software and sample preparation protocols for the identification and classification of microorganisms are currently available from several manufacturers (*63*, *75–78*). Consequently, more detailed studies are being conducted to optimize sample preparation for a wide range of microorganisms, and results obtained will add to a better understanding of the complex interplay of the variables involved in the sample preparation of microorganisms for MALDI-MS analysis. Finally, it is worth noting the findings by an inter-laboratory study (*79*) that the quality and reproducibility of profiles also depend on the model and manufacturer of the MALDI-MS instrument (*vide infra*). As such, optimized sample preparation conditions derived from using one type of instrument may not be optimal in another one.

MALDI Matrices for Microorganism Analysis

The MALDI process (*12*, *80*, *81*) is highly dependent on the nature of the analyte (peptide, carbohydrate, oligonucleotide, etc) and this in turn dictates the matrix used. Suitability of a compound as a MALDI matrix depends on factors like photochemical stability (in the wavelength range of the laser used), low vapor pressure and a high molar absorptivity at the laser wavelength when in the solid phase (*82*). Studies involving MALDI-MS for the detection of protein biomarkers from microorganisms have used several matrices including α-cyano-4-hydroxycinnamic acid (αCHCA), sinapinic acid (SA), 2,4-dihydroxybenzoic acid (DHB), ferrulic acid (FA) and 5-chloro-2-mercaptobenzothiazole (CMBT). Some commercial MALDI-MS based bacteria identification systems recommend αCHCA for the analysis of Gram-negative bacteria, while CMBT for the analysis of Gram-positive bacteria (*74*). It's worth highlighting a recent report by Fagerquist and co-workers (*83*) showing the formation of a covalent attachment *via* a thiol ester between the SA matrix and cysteine-containing protein biomarkers from bacterial cell lysates of *E. coli*, and as a result, care must be taken when comparing mass spectral fingerprints obtained with SA. The matrix αCHCA does have advantages in terms of a wide range of biomarkers that can be detected (peptides and small proteins), high reproducibility of profiles and formation of uniform crystals which facilitates automated data acquisition (*84*). However, it is important to keep in mind that optimum matrix performance is tied in a complex interplay with variables like the type of solvent(s) used to prepare the matrix (*85*), the solvent used to treat the microorganism being analyzed, the deposition method and the target biomarker. As a result, is not unusual to find in the literature different matrices being described as the "best" choice for microbial fingerprinting by MALDI-MS (*86–88*). That is, one type of matrix may perform better than another one only under certain experimental conditions and *vice versa*. Hence, it is highly advisable that a direct comparison between available matrices be performed under the particular set of conditions chosen to prepare the microbial sample (*vide infra*), using the same instrumentation and followed by a systematic optimization of the matrix/bacteria ratio used for MALDI-MS (*89*).

Factors in Sample Preparation Affecting the Reproducibility of MALDI-MS Analysis of Microorganisms

Several steps, carried out in a consistent and reproducible manner, are necessary for successful sample preparation of bacteria for MALDI-MS fingerprinting analysis. Important decisions must also be made in terms of growing and sampling of the microorganism(s) prior its final sample preparation for MALDI-MS analysis:

- Bacteria growth conditions (media, age of culture, sampling)
- Bacteria concentration
- Bacteria suspension solvent
- Bacteria pre-treatment (*e.g.*, inactivation, washing step, etc.)
- MALDI matrix
- Matrix solvent
- Matrix/bacteria ratio
- Solvent additives (e.g., acids, desalting agents, etc.)
- Method for sample deposition on the plate

Several studies have been conducted in order to establish a clear cause and effect of these experimental variables on the final quality (signal-to-noise ratio, number of signals detected) and selectivity (conserved versus unique biomarker signals) of the microbial mass spectral fingerprints. Also, not part of the sample preparation process, but a factor determining the reproducibility of the MALDI-mass profiles is the type and manufacturer of the MALDI-MS instrument.

Musser and co-workers (*90*) studied sample preparation as well as instrumentation factors that affect the reproducibility of the profiles obtained. Studied experimental variables related to bacterial sample preparation included the type of matrix, the matrix solvent, concentration of acid in the matrix solution and the concentration of cells in the matrix. In addition, two methods for sample sterilization, cells boiled in water or suspended in 70% ethanol, were tested. It was reported that an increased number of protein signals were observed for cells boiled in water, but the use of 70% ethanol was deemed more practical for the sterilization of pathogens and/or environmental samples (excluding spores, *vide infra*). Investigators also studied the effect of steps to remove salts (and other non-protein cell components) and found them to have little to no effect on the resulting quality of the bacterial mass profiles. Their final optimized protocol includes sterilization with 70% ethanol, αCHCA matrix in 50% acetonitrile/water with 2.5% trifluoro acetic acid (TFA), coupled to a serial dilution of the bacterial sample (to optimize matrix:cell concentration ratio when the microbial sample cell density is not known).

Wunschel and co-workers (*91*, *92*) studied the effects of bacterial growth conditions, mainly growth media pH, growth rate and temperature, on the MALDI-mass spectral bacterial profiles. Results showed that differences in growth rate created the largest variations in mass spectral profiles, and results were consistent with previously reported results by Arnold *et al* (*67*) where mass spectral fingerprints varied due to the age of the bacterial culture. However, with the implementation of their proprietary algorithm for mass spectral comparison (*93*), accurate classification was achieved despite of the variances observed. Looking also at variables affecting the reproducibility of MALDI-mass spectra of bacteria, Harrington and co-workers implemented a combination of analysis of variance and principal component analysis (ANOVA-PCA) to differentiate four strains of *E. coli* at different growth stages (*94*). Their data analysis showed that the greatest variance was due to bacterial sample age (mostly in samples < 24 hours growth), while instrument drift had no significant contribution to the variance in the MALDI-mass spectra of bacteria.

In an effort to better understand the sources of variations in bacteria mass spectral profiles obtained by MALDI-MS analysis, an inter-laboratory study was conducted between laboratories at the Pacific Northwest National Laboratory (PNNL, Richland, WA), Johns Hopkins University Applied Physics Laboratory (JHUAPL, Laurel, MD) and the National Institute of Standards and Technology (NIST, Gaithersburg, MD) (*79*). Three different MALDI-MS instruments were used (from two manufacturers), all operating with a nitrogen laser at 337.1 nm. Sample analyses were coordinated over a 3 day period between the laboratories and the data analyzed at PNNL. A single batch of bacterial sample (*E. coli*) as a water suspension and a single batch solution of matrix (SA, 20 mg/mL in 0.1% TFA 70% acetonitrile/water) was prepared and divided among the 3 laboratories. Even dilution solutions, external calibration standard solutions and vials were provided to minimize variability between the participating laboratories. Regardless of the care taken in using the same bacterial sample and reagents, profiles obtained in this study showed marked differences in peak intensities and in the number of peaks present, with nine ions being conserved in the mass spectra from all three laboratories. With the appropriate algorithm (*93*), identification was possible even when using a single bacteria mass spectral library against mass spectra obtained in different instruments. However, this study stressed the experimental variability introduced by the use of different MALDI-MS instruments, even after great care was taken in the standardization of the protocol, bacterial sample and reagents.

Sample preparation in MALDI-MS of intact bacteria can also be used to enhance the number and intensity of the protein signals observed. Signal enhancement in the MALDI-MS analysis of bacterial spores was reported in a study by Horneffer and co-workers (*95*, *96*) by treating aqueous bacterial suspensions to a 120 °C "wet heat-treatment" for 10-20 min. Their final protocol for the analysis of spores involved wet heat-treatment of the aqueous spore suspension, deposition of the treated spores onto the MALDI plate, dried and followed by application of SA matrix (70% acetonitrile/0.1% TFA aqueous solution).

Efforts to develop a "universal sample preparation method" for bacteria sample preparation for MALDI-MS analysis was also reported by Yang and co-workers (*88*). Their study evaluated parameters that affected reproducibility and profile quality (signal intensity and number of proteins detected) and in doing so developed an optimized sample preparation protocol. The protocol was evaluated with 9 bacterial species including both Gram-negative and Gram-positive bacteria. In the final protocol, the bacteria sample was twice washed, vortexed, centrifuged, re-suspended in 0.1% TFA and deposited onto the MALDI probe followed by application of matrix solution. The matrix solution consisted of αCHCA in a solution of 1:1:1 acetonitrile:methanol:water solution with 0.1% formic acid and 0.01 M 18-crown-6 (as a desalting agent). Due to the limited number of bacteria samples evaluated in this study, it is difficult to ascertain the universal applicability of this method on a wide range of bacteria species. In fact, even though spore-producing bacteria were tested, the method was not evaluated with actual bacterial spores, oocyst (*97*), fungi (*98*, *99*) or viruses (*100*), and its optimal performance in treating a wide range of microbial samples remains to be demonstrated.

Sample Treatment for the Safe Handling and Analysis of Microbial Pathogens by MALDI-MS

Microorganism sterilization should also be considered if the possibility of dealing with pathogenic samples exists, as it is the case in clinical, biodefense and forensic scenarios. Moreover, the MALDI-MS instruments may not be physically located within the confines of a BSL-3 microbiology laboratory, which creates logistical issues in the transfer and transport of potentially active pathogenic agents outside the BSL-3 facility, even if mixed with matrix and spotted onto the MALDI probe. As a result, several studies have incorporated an inactivation step into the bacteria-MALDI sample preparation protocol.

One approach by Williams *et al* (*90*), mentioned earlier in this chapter, treated cells in boiling water for 5 min or with 70% ethanol. Both methods were found effective in sterilizing the sample as checked by re-inoculation in growth media. Interestingly, both approaches were found not only to sterilize the sample, but also to increase the number of protein signals as in the wet heat-treatment mentioned in the previous section. However, the authors warn that the 70% ethanol treatment may be ineffective in killing spores. Addressing this particular issue, Lasch and co-workers developed an inactivation protocol for highly pathogenic (BSL-3) bacteria and spores that is compatible with MALDI-MS analysis (*101*, *102*). Their thorough approach involves the treatment of the bacterial sample with 80% TFA for 30 min (at room temperature) for complete inactivation of vegetative cells. For spore samples an additional centrifugation step (16,000 g for 20 min) followed by supernatant filtration (0.22 μm TFA resistant filter) assured complete inactivation. The resulting suspension/solution can be safely transported outside a BSL-3 laboratory for subsequent MALDI-MS (or ESI-MS) analysis, which involved sample dilution (1:9 with water) and mixing a 5 μL aliquot (1:1 (v/v)) with the matrix solution (αCHCA in 2:1 (v/v) acetonitrile:0.3% TFA). Profiles obtained after the inactivation process, when compared to controls, preserved strong biomarker signals in both vegetative cells and sporulated samples.

Microbial Sample Preparation for Clinical Applications of MALDI-MS Profiling

The need for controlled microorganism sample growth and pure cultures in MALDI-MS profiling of microorganisms makes this technique highly compatible within the clinical laboratory workflow. A controlled microbial sample preparation (sampling and growth) is part of the existing protocol in the clinical laboratory, and as a result, the incorporation of MALDI-MS bacteria profiling into the clinical protocol could potentially replace phenotypic and biochemical test currently used. In fact, several studies have made direct comparisons between MALDI-MS (using commercially available microbial identification systems) and established biochemical tests (*78*, *103*, *104*) with encouraging results. In one study, Seng (*103*) and co-workers correctly identified 95.4% of

1660 bacterial isolates analyzed by MALDI-MS. In another interlaboratory study Mellmann and co-workers (*105*) eight international laboratories analyzed 60 blind-coded nonfermenting bacteria samples and achieved an impressive 98.8% reproducibility, with only 6 out of 480 samples misidentified (mostly due to interchanges and contamination). This, of course, was achieved through strict control of experimental conditions: all laboratories used the same MALDI-MS instrument, identification software and database, the same batch of bacteria samples and all followed a "sample cultivation and preparation guide" and "result reporting guide".

In general, sample preparation for MALDI-MS profiling of clinical samples begins with inoculation of growth (selective) media with a biological fluid sample from a patient, followed by isolation of single colonies after a pre-determined incubation period. Growth media can be either broth, usually requiring an extra washing step, or agar plates, which can be sampled with a sterile loop and smeared directly onto the MALDI plate or suspended in solution. This final microbial suspension was directly deposited onto the MALDI plate. In some instances, an extra inactivation/wash step (addition of ethanol) and/or cell disruption step (addition of TFA) was added before addition of the MALDI matrix. Another report used an ethanol treatment (*68*) before addition of matrix for samples smeared directly onto the MALDI plate. The most common matrix used for these clinical studies was αCHCA in 50% acetonitrile:2.5% aqueous TFA. An overview (not comprehensive) of the sample preparation protocols of several clinical studies using MALDI-MS profiling for bacteria identification is presented in Table 1.

Even though a high level of reproducibility has been achieved, accuracy of the bacteria identification *via* MALDI-MS profiling in a clinical setting will continue to be studied by many laboratories. Advances in the standardization of the MALDI sample preparation protocol(s) for the wide range of microbial species encountered in the clinical setting (and different types of MALDI-MS instruments in clinical laboratories) could lead to the integration of MALDI-MS as a rapid, cost-effective (*103*) and routine method for the identification of clinical isolates.

Sample Preparation for MALD-MS Profiling of Microbial Mixtures and Environmental Samples

The application of MALDI-MS for the rapid detection and identification of microorganisms in the environment remains a challenging process. This stems from the fact that in environmental samples like food or recreational waters a microbial mixture may be present, the target microorganism may be at a low level and/or the chemical background from the sample matrix may interfere with the analysis. As a result, in order to obtain a MALDI-MS profiling of a pure microbial culture from an environmental sample, strategies must be incorporated into the sample preparation protocol for the separation and enrichment of the target microorganism(s) prior to MALDI-MS analysis. This however, is only true for MALDI-MS microorganism profiling, as the microorganism mass spectrum will be compared to a library of standard microbial fingerprints. These strategies usually involve standard microbiology protocols involving selective media and other enrichment strategies in order to obtain pure cultures of the microorganism

Table 1. Selected clinical-related applications of MALDI-MS bacteria profiling

Microorganism/ Growth media	*Sample deposited on plate*	*Wash/ extraction*	*Cell pre-treatment*	*Matrix (solvent)*	*Ref*
1,660 clinical isolates/agar & broth	Single colony smear	-	-	αCHCA (50% ACN:2.5% TFA)	(*103*)
432 clinical isolates /blood culture broth	Cell suspension (water)	Water wash	10% TFA (to cell pellet)	DHB (30% ACN:0.1% TFA)	(*106*)
327 clinical isolates/agar	Single colony smear or protein extract	Ethanol wash 1:1 70% FA:ACN	-	αCHCA (50% ACN:2.5% TFA)	(*107*)
304 aerobic anaerobic blood cultures/broth	Protein extract	Water and ethanol washes/ 1:1 70% FA:ACN	-	αCHCA (50% ACN:2.5% TFA)	(*104*)
122 isolates/ blood culture	Protein extract	Water wash/ 1:1 70% FA:ACN	Ammonium chloride lysing	αCHCA (50% ACN:2.5% TFA)	(*108*)
559 Bacilli from cystic fibrosis patients/agar	Cell suspension (water)	-	Absolute ethanol (on plate)	DHB (30% ACN:0.1% TFA)	(*109*)
1,371 clinical isolates/	Single colony smear	-	-	αCHCA (50% ACN:2.5% TFA)	(*110*)
75 strains *Burkholderia cepacia* complex/ agar	Cell suspension (in 0.1 % TFA) pre-mixed with matrix solution	-	0.1 % TFA (initial cell suspension)	αCHCA (49%ACN: 49% isopropanol: 0.1% TFA)	(*111*)
Arcobacter, Helicobacter, Campylobacter/ agar	Protein extract	Water wash and ethanol wash-inactivation/ 1:1 70% FA:ACN	-	αCHCA (50% ACN:2.5% TFA)	(*112*)
21 *Legionella* species/agar	Single colony smear	-	-	αCHCA (50% ACN:2.5% TFA)	(*113*)
17 *Bartonella* species/agar	Cell suspension (water)	-	-	αCHCA (50% ACN:2.5% TFA)	(*114*)

for MALDI-MS analysis. In situations where growing the microbial sample is not feasible or the time scale for the analysis needs to be near real-time (*i.e.*, less than 5 min total analysis time), the strategy has involved the use of affinity probes (*115*, *116*) or bioactive surfaces (*117–120*) for the specific enrichment of the target microorganism prior to MALDI-MS analysis. Some examples of recent developments in both areas of research will be discussed next.

Several investigators have implemented MALDI-MS for the characterization and development of unique mass spectral fingerprints for several food pathogens including *Vibrio parahaemolyticus* (*121*), *Listeria* species (*122*) and amine-producing bacteria (*123*). The analysis of food samples is challenging since the sample matrix may contain a very complex mixture of proteins, carbohydrates, salts, fats and additives. Moreover, the target microorganism(s) may be at levels below the analytical detection limit of the method, making their detection difficult. Several protocols have been developed that address this challenging sample composition. In work presented by Parisi and co-workers (*124*) a protocol was developed for the sampling, enrichment and MALDI-MS profiling of *Yersinia enterocolitica* in spiked bovine meat samples. The spiked meat sample (~25 g, *Y. enterocolitica* <10 CFU/g) was first homogenized and incubated with buffered peptone water (*125*) for 2 hours. A small aliquot of this solution was then incubated with *Yersinia* enrichment broth for an additional 6 hours. After this second incubation period, the sample was analyzed by MALDI-MS. Authors reported no false-negative or false-positive results using the spiked samples, which alludes to the applicability of this method for the identification of bacteria in real samples. Calo-Mata and co-workers (*126*) reported on the detection and identification by MALDI-MS profiling of spoiling and pathogenic Gram-negative bacteria previously isolated from seafood samples. A standard protocol for isolating and enriching bacteria was applied before samples were analyzed by MALDI-MS. Sample preparation for MALDI-MS involved a simple protein extraction step with a solution of 50% acetonitrile:1% aqueous TFA followed by centrifugation. The supernatant was analyzed by MALDI-MS using αCHCA. With this method, the authors obtained good correlation between MALDI-MS profiles and clustering of the same microorganisms *via* phylogenetic analysis by 16S rRNA. In both studies, the sample was processed utilizing established microbiology protocols, with the resulting enriched and pure sample being analyzed by MALDI-MS. This approach allowed the comparison of standard microbial mass fingerprints with those obtained from the MALDI-MS analysis of the unknown sample.

Affinity probes are widely used in the MS for the analysis of low level target analytes in biological samples, as it is the case for the detection of phosphoproteins (*127*). This approach has also been applied to the detection of microorganisms by MADLI-MS profiling for the analysis of microbial mixtures or microorganisms in complex matrices (*e.g.*, biological fluids) (*115*, *116*, *120*). Recent work by Y.-C. Chen and co-workers applied functionalized nanoparticles (NPs) to isolate and enrich bacteria from spiked samples for subsequent analysis by MALDI-MS. Nanoparticles have been coated with human immunoglobin (IgG) (*128*), vancomycin (*129*) and pigeon ovalbumin (*130*) to enrich bacteria from mixtures and spiked urine samples. The vancomycin functionalized NPs showed enhanced specificity towards Gram-positive bacteria and were used to isolate *Staphylococcus saprophyticus* from urine samples. MALDI-MS profiles of the isolated bacteria in spiked urine samples showed distinctive species-specific bacterial signals from samples with a concentration of $7x10^5$ cfu/mL.

The detection of microorganisms in recreational waters and water supplies is a challenging problem since the level of detection needed is well below that of conventional MALDI-MS profiling. As a result, this requires a sample preparation step incorporating an enrichment process and/or the processing of large volumes of sample in order to collect sufficient cells for a successful MALDI-MS analysis. Addressing this problem, Guo's group applied the functionalized NPs approach described earlier to isolate bacteria from spiked water samples and obtain MALDI-MS profiles for their detection and identification. Using commercially available anion-exchange superparamagnetic NPs, non-specific and broad spectrum isolation of Gram-positive and Gram-negative bacteria was achieved (*131*). Incorporation of an initial membrane filtration step (0.45 μm cellulose nitrate) followed by functionalized NP trapping (*132*), bacteria detection limits near 10^3 cfu/mL were obtained for 2.0 L water samples with a total analysis time of 2 hours (*133*).

In all these studies, the resulting MALDI-MS bacterial profile was similar to the standard mass spectrum; however, with many background signals also detected. The current limited specificity that can be achieved with affinity probes and the presence of background peaks (*i.e.*, cross-reactivity) make this approach challenging to implement in a routine laboratory workflow. As a result, the success of this approach depends heavily on the development of new specific affinity probe(s) towards the target microorganism. Also, the incorporation of proteomic-based approaches could circumvent the need for specific isolation and enrichment (*vide infra*).

Sample Preparation for Proteomic-Based Microorganism Identification

In proteome-based microorganism analysis (*59*, *134*), the identification is achieved by matching tandem MS (MS/MS) data of a peptide (bottom-up approach) or protein (top-down approach) to a database. In general, for a top-down approach to microorganism identification and utilizing LC-ESI-MS instrumentation, the microbial sample needs to be lysed and proteins extracted/purified. On the other hand, performing the analysis utilizing MALDI-MS instrumentation, the sample preparation protocol is simplified considerably and is analogous to the sample preparation steps in MALDI-MS profiling of microorganisms (*vide supra*). If analyzing *via* a bottom-up approach, a site-specific digestion of the proteins into peptides is required for successful analysis (*135*, *136*). This site-specific digestion is normally carried out with proteolytic enzymes (*62*, *137*), the most commonly used being trypsin. Hence, the high level of specificity obtained from this analysis comes at the cost of an additional sample preparation step with biological reagents (*i.e.*, enzymes). Consequently, efforts have been placed toward the development of approaches to speed up this critical digestion step in the sample preparation protocol by either accelerating tryptic digestions with microwave radiation heating (*138*) or implementing a non-enzymatic protein digestion step also accelerated by microwave heating (*139*, *140*). The obvious advantage of the proteomic-based protocol is its independence on yielding profiles with reproducible peak heights and ratios, and as a result sample preparation is not as critical in determining the successful identification of biomarker protein(s) (and thereby the identity of the microorganisms) with either ESI-MS/MS or MALDI-MS/MS.

Sample Preparation for Bottom-Up Proteomic Identification of Microorganisms

The sample preparation for bottom-up proteomic microorganism identification is analogous to standard proteomic analyses performed on a wide range of biological systems, and early work in this area implemented enzymatic digestion of the intact microorganism with trypsin. This assay can analyze mixtures and thus microorganism fractionation is not required. When implementing MALDI-MS, the digestion was carried out on the MALDI plate surface (often reffered to as on-probe or on-target digestion) maintained inside a humidity chamber for 20 min (*137*, *141*). The approach was also applied to mixtures of bacteria that were digested on probe with immobilized trypsin for 20 min (*142*). Trypsin immobilization reduced the formation of autohydrolysis products and allowed for an increase concentration of enzyme during the digestion process. Recently, a similar approach was applied to aerosolized bacteria collected directly onto the MALDI plate. The collected bacteria was subsequently digested on-target and detected *via* a peptide mass fingerprinting database search (*143*). Affinity probes have also been implemented to sample and preconcentrate microbial sample, bypassing the culturing step, prior to protein digestion and MALDI-MS/MS. Naoparticles (NPs) coated with titania (TiO_2) were used to capture and isolate the bacteria. Typtic digestion was performed on the isolated bacteria and enhanced by microwave heating (1 min digestion) prior to MALDI-MS/MS analysis (*144*). With this approach, the authors were able to analyze 10^4 CFU/mL bacteria suspensions with a 15 min isolation period.

When the MS analysis is performed using ESI as the ionization source, the resulting complex mixture of peptides requires the implementation of liquid chromatography in order to avoid suppression effects (*10*, *145*). In this approach (*146*, *147*), the bacteria sample is prepared following a standard proteomic approach involving cell lysis, protein digestion (with trypsin) and sample cleanup (precipitation, washes, centrifugation, etc.), making this approach impractical as a rapid identification technique. However, it is well documented that implementation of a proteomic-based approach using LC-ESI-MS to bacteria identification leads to the detection of more and larger molar mass protein biomarkers than when using MALDI-MS detection (*148*, *149*), which may find useful applications in confirmatory analyses.

The use of enzymes requires special handling in terms of buffers and temperature, and storage conditions (-20 °C), as a result, the shelf life of these reagents may be limited and their use in field-portable instruments difficult to implement. To circumvent some of these problems, methods based on site-specific chemical cleavage (*150–152*) of the protein backbone have been implemented in proteomic approaches. The cleavage at aspartic acid, induced by acid hydrolysis and heat, has been useful in proteomic approaches (*153*) since the reagent used is compatible with either ESI or MALDI-MS detection. Microwave heating was soon incorporated into the assay in order to accelerate the acid hydrolysis process, making it possible to site-specifically digest proteins in less than 5 min and without any additional reagent (*139*). Incorporation of this method into a proteomic-based bacteria identification workflow soon followed for the rapid analysis of spores (*140*) and intact human adenovirus (*154*). Development of a microwave flow cell allowed the online digestion and ESI-MS/MS analysis of digest products of *E. coli* proteins without sample handling (*155*).

Sample Preparation for Top-Down Proteomic Identification of Microorganisms

The top-down proteomic approach to identify proteins relies on the fragmentation of an intact protein in a tandem MS experiment to yield a partial amino acid sequence and/or peptide fragments. As in the bottom-up proteomic approach, the assay can analyze mixtures and thus microorganism fractionation is not required. Protein fragmentation in the MS can be achieved either by Collision Induced Dissociation (CID) (*156*, *157*), Laser Induced Disoociation (LID) (*158*), Electron Capture Dissociation (ECD) (*159*, *160*) or Electron Transfer Dissociation (ETD) (*161*). In top-down proteomic strategies using ESI to generate ions, protocols are simpler since the protein digestion step is omitted, with only cell lysis and protein extraction/purification remaining as key steps in this assay (*162–164*). However, any advantage in terms of rapid sample preparation time saved by omitting the protein digestion step is precluded by the need to include a LC separation step in order to avoid signal suppression during ESI. When MALDI is used as the ionization source for top-down proteomic approaches, whole cell MALDI analysis can be implemented using the same protocols outlined for MALDI-MS profiling (*vide supra*). Demirev and coworkers (*158*) analyzed intact spores by MALDI-MS/MS (Tof-Tof) which incorporated an on-probe 10% formic acid treatment in order to facilitate SASP extractions from spores (*165*). Fequerquist and coworkers (*166*) on the other hand chose to perform a protein extraction prior to MALDI-MS/MS (ToF-ToF) analysis by suspending bacterial cells in a solution of 67% water, 33% acetonitrile and 0.1% TFA with 0.1 mm zirconia/silica beads. The sample was bead-beated for 1 min, centrifuged and the supernatant used for MALDI-MS/MS analysis. Overall, these approaches have the potential to exploit the ease of intact cell sample preparation of MALDI with the high specificity of top-down proteomics.

Conclusions and Future Trends

From the survey presented here, "*Rapid microorganism identification by MS*" in the context of its application may require a sample preparation step that ranges from less than 5 minutes, as is the case in biodefense applications, to hours, as implemented in environmental applications (this analysis time does *not* include initial sample growth period). Advances in the sample preparation protocol are intimately tied to advances in MS instrumentation and bioinformatic tools to process the mass spectral data. Top-down proteomic approaches require less "wet chemistry" sample preparation and are not profile-based, and as a result it is expected that their implementation for the analysis and identification of microorganism in complex mixtures will increase. The usefulness and ruggedness of MALDI as an intact cell sample preparation step to produce high molar mass protein biomarkers will continue to play a crucial role in this endeavor. The implementation of MALDI with top-down proteomics will require further advances in strategies involving ECD, ETD, ion/ion and/or ion/molecule reactions (*167*) to increase the efficiency of fragmentation of the large protein ions generated by MALDI.

A fully automated and rapid biodefense application of MS calls for the elimination of the "man-in-the-loop" of the sample preparation protocol. As such, the application of strategies implementing automated sample preparation systems based on flow-injection-analysis with microfluidics should succeed at this task. Further developments in the application of Ambient-MS techniques for the analysis of microorganisms should be expanded towards the detection of protein/peptide biomarkers, which will allow the implementation of proteomic-based approaches and the analysis of microbial mixtures.

Finally, as MALDI-MS bacteria profiling is bound to play a key role in the clinical laboratory (*168*), improvements and development of a standard bacteria sample preparation protocol(s) to yield reproducible mass spectral fingerprints will undoubtedly require further interlaboratory studies and the incorporation of automated sample preparation steps.

References

1. Anhalt, J. P.; Fenselau, C. Identification of bacteria using mass spectrometry. *Anal. Chem.* **1975**, *47* (2), 219–225.
2. DeLuca, S.; Sarver, E. W.; Harrington, P. d. B.; Voorhees, K. J. Direct analysis of bacterial fatty acids by Curie-point pyrolysis tandem mass spectrometry. *Anal. Chem.* **1990**, *62* (14), 1465–72.
3. Dworzanski, J. P.; Berwald, L.; Meuzelaar, H. L. C. Pyrolytic methylation-gas chromatography of whole bacterial cells for rapid profiling of cellular fatty acids. *Appl. Environ. Microbiol.* **1990**, *56* (6), 1717–24.
4. Heller, D. N.; Cotter, R. J.; Fenselau, C.; Uy, O. M. Profiling of bacteria by fast atom bombardment mass spectrometry. *Anal. Chem.* **1987**, *59* (23), 2806–9.
5. Cole, M. J.; Enke, C. G. Direct determination of phospholipid structures in microorganisms by fast atom bombardment triple quadrupole mass spectrometry. *Anal. Chem.* **1991**, *63* (10), 1032–8.
6. Fenn, J. B.; Mann, M.; Meng, C. K.; Wong, S. F.; Whitehouse, C. M. Electrospray ionization for mass spectrometry of large biomolecules. *Science* **1989**, *246* (4926), 64–71.
7. Karas, M.; Hillenkamp, F. Laser desorption ionization of proteins with molecular masses exceeding 10,000 daltons. *Anal. Chem.* **1988**, *60* (20), 2299–301.

8. Fenn, J. B.; Mann, M.; Meng, C. K.; Wong, S. F.; Whitehouse, C. M. Electrospray ionization-principles and practice. *Mass Spectrom. Rev.* **1990**, *9* (1), 37–70.
9. Gibson, G. T. T.; Mugo, S. M.; Oleschuk, R. D. Nanoelectrospray emitters: trends and perspective. *Mass Spectrom. Rev.* **2009**, *28* (6), 918–936.
10. Kebarle, P.; Verkerk, U. H. Electrospray: from ions in solution to ions in the gas phase, what we know now. *Mass Spectrom. Rev.* **2009**, *28* (6), 898–917.
11. Hillenkamp, F.; Karas, M. Matrix-assisted laser desorption/ionisation, an experience. *Int. J. Mass Spectrom.* **2000**, *200* (1/3), 71–77.
12. Zenobi, R.; Knochenmuss, R. Ion formation in MALDI mass spectrometry. *Mass Spectrom. Rev.* **1998**, *17* (5), 337–366.
13. Zenobi, R.; Breuker, K.; Knochenmuss, R.; Lehmann, E.; Stevenson, E. Fundamentals and applications of MALDI mass spectrometry. *Adv. Mass Spectrom.* **2001**, *15*, 143–149.
14. Reyzer, M. L.; Caprioli, R. M. In situ proteomics. *Mod. Drug Discovery* **2002**, *5* (10), 45–46, 48.
15. Dalluge, J. J. Matrix-assisted laser desorption ionization-mass spectrometry. *Anal. Bioanal. Chem.* **2002**, *372* (1), 18–19.
16. Takats, Z.; Wiseman, J. M.; Gologan, B.; Cooks, R. G. Mass Spectrometry Sampling Under Ambient Conditions with Desorption Electrospray Ionization. *Science* **2004**, *306* (5695), 471–473.
17. Cody, R. B.; Laramee, J. A.; Durst, H. D. Versatile new ion source for the analysis of materials in open air under ambient Conditions. *Anal. Chem.* **2005**, *77* (8), 2297–302.
18. Takats, Z.; Wiseman, J. M.; Cooks, R. G. Ambient mass spectrometry using desorption electrospray ionization (DESI): instrumentation, mechanisms and applications in forensics, chemistry, and biology. *J. Mass Spectrom.* **2005**, *40* (10), 1261–75.
19. Cooks, R. G.; Ouyang, Z.; Takats, Z.; Wiseman, J. M. Detection Technologies. Ambient mass spectrometry. *Science* **2006**, *311* (5767), 1566–70.
20. Venter, A.; Nefliu, M.; Cooks, R. G. Ambient desorption ionization mass spectrometry. *TrAC, Trends Anal. Chem.* **2008**, *27* (4), 284–290.
21. Chen, H.; Gamez, G.; Zenobi, R. What can we learn from ambient ionization techniques? *J. Am. Soc. Mass Spectrom.* **2009**, *20* (11), 1947–63.
22. Green, F. M.; Salter, T. L.; Stokes, P.; Gilmore, I. S.; O'Connor, G. Ambient mass spectrometry: advances and applications in forensics. *Surf. Interface Anal.* **2010**, *42* (5), 347–357.
23. Takats, Z.; Wiseman, J. M.; Cooks, R. G. Ambient mass spectrometry using desorption electrospray ionization (DESI): instrumentation, mechanisms and applications in forensics, chemistry, and biology. *J. Mass Spectrom.* **2005**, *40* (10), 1261–75.
24. Song, Y.; Talaty, N.; Tao, W. A.; Pan, Z.; Cooks, R. G. Rapid ambient mass spectrometric profiling of intact, untreated bacteria using desorption electrospray ionization. *Chem. Commun.* **2007**, *1* (1), 61–63.

25. Meetani, M. A.; Shin, Y.-S.; Zhang, S.; Mayer, R.; Basile, F. Desorption electrospray ionization mass spectrometry of intact bacteria. *J. Mass Spectrom.* **2007**, *42* (9), 1186–1193.
26. Zhang, J. I.; Talaty, N.; Costa, A. B.; Xia, Y.; Tao, W. A.; Bell, R.; Callahan, J. H.; Cooks, R. G., Rapid direct lipid profiling of bacteria using desorption electrospray ionization mass spectrometry. *Int. J. Mass Spectrom.* 2010, In Press, Corrected Proof.
27. Voorhees, K. J.; Durfee, S. L.; Updegraff, D. M. Identification of diverse bacteria grown under diverse conditions using pyrolysis-mass spectrometry. *J. Microbiol. Methods* **1988**, *8*, 315–325.
28. DeLuca, S. J.; Sarver, E. W.; Voorhees, K. J. Direct analysis of bacterial glycerides by Curie-point pyrolysis mass spectrometry. *J. Anal. Appl. Pyrolysis* **1992**, *23*, 1–14.
29. Basile, F.; Beverly, M. B.; Hadfield, T. L.; Voorhees, K. J. Pathogenic bacteria: their detection and differentiation by rapid lipid profiling with pyrolysis mass spectrometry. *Trends Anal. Chem.* **1998**, *17* (2), 95–109.
30. Song, Y.; Talaty, N.; Datsenko, K.; Wanner, B. L.; Cooks, R. G. In vivo recognition of Bacillus subtilis by desorption electrospray ionization mass spectrometry (DESI-MS). *Analyst* **2009**, *134* (5), 838–841.
31. Venter, A.; Cooks, R. G. Desorption Electrospray Ionization in a Small Pressure-Tight Enclosure. *Anal. Chem.* **2007**, *79* (16), 6398–6403.
32. Mulligan, C. C.; Talaty, N.; Cooks, R. G. Desorption electrospray ionization with a portable mass spectrometer: in situ analysis of ambient surfaces. *Chem. Commun.* **2006** (16), 1709–11.
33. Cotte-Rodriguez, I.; Cooks, R. G. Non-proximate detection of explosives and chemical warfare agent simulants by desorption electrospray ionization mass spectrometry. *Chem. Commun.* **2006**, *28* (28), 2968–70.
34. Song, Y.; Cooks, R. G. Reactive desorption electrospray ionization for selective detection of the hydrolysis products of phosphonate esters. *J. Mass Spectrom.* **2007**, *42* (8), 1086–1092.
35. Zhang, Y.; Chen, H. Detection of saccharides by reactive desorption electrospray ionization (DESI) using modified phenylboronic acids. *Int. J. Mass Spectrom.* **2010**, *289* (2−3), 98–107.
36. Beverly, M. B.; Basile, F.; Voorhees, K. J.; Hadfield, T. L. A rapid approach for the detection of dipicolinic acid in bacterial spores using pyrolysis/mass spectrometry. *Rapid Commun. Mass Spectrom.* **1996**, *10* (4), 455–58.
37. Fox, A.; Black, G.; Fox, K.; Wunschel, D. Identification and detection of bacteria: electrospray MS-MS versus derivatization/GC-MS. *Proc. ERDEC Sci. Conf. Chem. Biol. Def. Res* **1996**.
38. White, D. C.; Lytle, C. A.; Gan, Y.-D. M.; Piceno, Y. M.; Wimpee, M. H.; Peacock, A. D.; Smith, C. A. Flash detection/identification of pathogens, bacterial spores and bioterrorism agent biomarkers from clinical and environmental matrices. *J. Microbiol. Methods* **2002**, *48* (2-3), 139–147.
39. Srivastava, A.; Pitesky, M. E.; Steele, P. T.; Tobias, H. J.; Fergenson, D. P.; Horn, J. M.; Russell, S. C.; Czerwieniec, G. A.; Lebrilla, C. B.; Gard, E. E.; Frank, M. Comprehensive assignment of mass spectral signatures from individual Bacillus atrophaeus spores in matrix-free laser

desorption/ionization bioaerosol mass spectrometry. *Anal. Chem.* **2005**, *77* (10), 3315–3323.

40. Murrell, W. G. Chemical Composition of Spores and Spore Structures. In *The Bacterial Spore*; Gould, G. W., Hurst, A., Eds.; Academic Press: London, 1969; Vol. I, pp 215−273.
41. Setlow, B.; Atluri, S.; Kitchel, R.; Koziol-Dube, K.; Setlow, P. Role of Dipicolinic Acid in Resistance and Stability of Spores of Bacillus subtilis with or without DNA-Protective alpha/beta-Type Small Acid-Soluble Proteins. *J. Bacteriol.* **2006**, *188* (11), 3740–3747.
42. Warth, A. D. Liquid Chromatographic Determination of Dipicolinic Acid from Bacterial Spores. *Appl. Environ. Microbiol.* **1979**, *38* (6), 1029–1033.
43. Snyder, A. P.; Thornton, S. N.; Dworzanski, J. P.; Meuzelaar, H. L. C. Detection of the picolinic acid biomarker in Bacillus spores using a potentially field-portable pyrolysis-gas chromatography-ion mobility spectrometry system. *Field Anal. Chem. Technol.* **1996**, *1* (1), 49–59.
44. Goodacre, R.; Shann, B.; Gilbert, R. J.; Timmins, E. M.; McGovern, A. C.; Alsberg, B. K.; Kell, D. B.; Logan, N. A. Detection of the Dipicolinic Acid Biomarker in Bacillus Spores Using Curie-Point Pyrolysis Mass Spectrometry and Fourier Transform Infrared Spectroscopy. *Anal. Chem.* **2000**, *72* (1), 119–127.
45. Li, Q.; Dasgupta, P. K.; Temkin, H.; Crawford, M. H.; Fischer, A. J.; Allerman, A. A.; Bogart, K. H. A.; Lee, S. R. Mid-ultraviolet light-emitting diode detects dipicolinic acid. *Appl. Spectrosc.* **2004**, *58* (11), 1360–1363.
46. Navarro, A. K.; Pena, A.; Perez-Guevara, F. Endospore dipicolinic acid detection during Bacillus thuringiensis culture. *Lett. Appl. Microbiol.* **2008**, *46* (2), 166–170.
47. Phillips, T. E.; Sample, J. L.; Scholl, P. F.; Miragliotta, J. The use of surface enhanced Raman scattering for the detection of dipicolinic acid on silver nanoparticles. *Mater. Res. Soc. Symp. Proc.* **2003**, *738* (Spatially Resolved Characterization of Local Phenomena in Materials and Nanostructures), 227–232.
48. Farquharson, S.; Inscore, F. E. Detection of invisible bacilli spores on surfaces using a portable SERS-based analyzer. *Int. J. High Speed Electron. Syst.* **2008**, *18* (2), 407–416.
49. Holzer, G.; Bourne, T. F.; Bertsch, W. Analysis of in situ methylated microbial fatty acid constituents by Curie-point pyrolysis-gas chromatography-mass spectrometry. *J. Chromatogr.* **1989**, *468*, 181–190.
50. Voorhees, K. J.; Basile, F.; Beverly, M. B.; Abbas-Hawks, C.; Hendricker, A.; Cody, R. B.; Hadfield, T. L. The use of biomarker compounds for the identification of bacteria by pyrolysis-mass spectrometry. *J. Anal. Appl. Pyrolysis* **1997**, *40−41*, 111–134.
51. Basile, F.; Zhang, S.; Shin, Y.-S.; Drolet, B. Atmospheric pressure-thermal desorption (AP-TD)/electrospray ionization-mass spectrometry for the rapid analysis of Bacillus spores. *Analyst* **2010**, *135* (4), 797–803.
52. Pierce, C. Y.; Barr, J. R.; Cody, R. B.; Massung, R. F.; Woolfitt, A. R.; Moura, H.; Thompson, H. A.; Fernandez, F. M. Ambient generation of fatty

acid methyl ester ions from bacterial whole cells by direct analysis in real time (DART) mass spectrometry. *Chem. Commun.* **2007** (8), 807–809.
53. Song, L.; Gibson, S. C.; Bhandari, D.; Cook, K. D.; Bartmess, J. E. Ionization Mechanism of Positive-Ion Direct Analysis in Real Time: A Transient Microenvironment Concept. *Anal. Chem.* **2009**, *81* (24), 10080–10088.
54. Weston, D. J. Ambient ionization mass spectrometry: current understanding of mechanistic theory; analytical performance and application areas. *Analyst* **2010**, *135* (4), 661–668.
55. Basile, F.; Beverly, M. B.; Abbas-Hawks, C.; Mowry, C. D.; Voorhees, K. J.; Hadfield, T. L. Direct Mass Spectrometric Analysis of in Situ Thermally Hydrolyzed and Methylated Lipids from Whole Bacterial Cells. *Anal. Chem.* **1998**, *70* (8), 1555–1562.
56. Barshick, S. A.; Wolf, D. A.; Vass, A. A. Differentiation of Microorganisms Based on Pyrolysis-Ion Trap Mass Spectrometry Using Chemical Ionization. *Anal. Chem.* **1999**, *71*, 633–641.
57. Shin, Y.-S.; Drolet, B.; Mayer, R.; Dolence, K.; Basile, F. Desorption Electrospray Ionization-Mass Spectrometry of Proteins. *Anal. Chem.* **2007**, *79* (9), 3514–3518.
58. Miao, Z.; Chen, H. Direct Analysis of Liquid Samples by Desorption Electrospray Ionization-Mass Spectrometry (DESI-MS). *J. Am. Soc. Mass Spectrom.* **2009**, *20* (1), 10–19.
59. Demirev, P. A.; Fenselau, C. Mass spectrometry in biodefense. *J. Mass Spectrom.* **2008**, *43* (11), 1441–1457.
60. Russell, S. C. Microorganism characterization by single particle mass spectrometry. *Mass Spectrom. Rev.* **2009**, *28* (2), 376–387.
61. Lay, J. O., Jr. MALDI-TOF mass spectrometry of bacteria. *Mass Spectrom. Rev.* **2001**, *20* (4), 172–94.
62. Fenselau, C.; Demirev, P. A. Characterization of intact microorganisms by MALDI mass spectrometry. *Mass Spectrom. Rev.* **2001**, *20* (4), 157–71.
63. Giebel, R.; Worden, C.; Rust, S. M.; Kleinheinz, G. T.; Robbins, M.; Sandrin, T. R. Microbial Fingerprinting using Matrix-Assisted Laser Desorption Ionization Time-Of-Flight Mass Spectrometry (MALDI-TOF MS): Applications and Challenges. In *Advances in Applied Microbiology*; Gadd, A. I. L. S. S. G. M., Ed.; Academic Press: 2010; Vol. 71, pp 149−184.
64. Holland, R. D.; Wilkes, J. G.; Rafii, F.; Sutherland, J. B.; Persons, C. C.; Voorhees, K. J.; Lay, J. O. Rapid identification of intact whole bacteria based on spectral patterns using matrix-assisted laser desorption/ionization with time-of-flight mass spectrometry. *Rapid Commun. Mass Spectrom* **1996**, *10* (10), 1227–1232.
65. Claydon, M. A.; Davey, S. N.; Edwards-Jones, V.; Gordon, D. B. The rapid identification of intact microorganisms using mass spectrometry. *Nat. Biotechnol.* **1996**, *14* (11), 1584–1586.
66. Krishnamurthy, T.; Ross, P. L.; Rajamani, U. Detection of pathogenic and non-pathogenic bacteria by matrix-assisted laser desorption/ionization time-of-flight mass spectrometry. *Rapid Commun. Mass Spectrom.* **1996**, *10* (8), 883–888.

67. Arnold, R. J.; Karty, J. A.; Ellington, A. D.; Reilly, J. P. Monitoring the growth of a bacteria culture by MALDI-MS of whole cells. *Anal. Chem.* **1999**, *71* (10), 1990–6.
68. Madonna, A. J.; Basile, F.; Ferrer, I.; Meetani, M. A.; Rees, J. C.; Voorhees, K. J. On-probe sample pretreatment for the detection of proteins above 15 KDa from whole cell bacteria by matrix-assisted laser desorption/ionization time-of-flight mass spectrometry. *Rapid Commun. Mass Spectrom.* **2000**, *14* (23), 2220–9.
69. Saenz, A. J.; Petersen, C. E.; Valentine, N. B.; Gantt, S. L.; Jarman, K. H.; Kingsley, M. T.; Wahl, K. L. Reproducibility of Matrix-assisted Laser Desorpton/Ionization Time-of-flight Mass Spectrometry for Replicate Bacterial Culture Analysis. *Rapid Commun. Mass Spectrom.* **1999**, *13*, 1580–1585.
70. Lay, J. O., Jr.; Holland, R. D. Rapid identification of bacteria based on spectral patterns using MALDI-TOFMS. *Methods Mol. Biol.* **2000**, *146*, 461–87.
71. Evason, D. J.; Claydon, M. A.; Gordon, D. B. Effects of ion mode and matrix additives in the identification of bacteria by intact cell mass spectrometry. *Rapid Commun. Mass Spectrom.* **2000**, *14*, 669–672.
72. Evason, D. J.; Claydon, M. A.; Gordon, D. B. Exploring the limits of bacterial identification by intact cell-mass spectrometry. *J. Am. Soc. Mass Spectrom.* **2001**, *12* (1), 49–54.
73. Bright, J. J.; Claydon, M. A.; Soufian, M.; Gordon, D. B. Rapid typing of bacteria using matrix-assisted laser desorption ionisation time-of-flight mass spectrometry and pattern recognition software. *J. Microbiol. Methods* **2002**, *48* (2-3), 127–38.
74. Dare, D., Rapid Bacterial Characterization and Identification by MALDI-TOF Mass Spectrometry. In *Advanced Techniques in Diagnostic Microbiology*; Tang, Y.-W.; Stratton, C. W., Eds. Springer: New York, 2006; pp 117−133.
75. Jackson, K. A.; Edwards-Jones, V.; Sutton, C. W.; Fox, A. J. Optimisation of intact cell MALDI method for fingerprinting of methicillin-resistant Staphylococcus aureus. *J. Microbiol. Methods* **2005**, *62* (3), 273–284.
76. Maier, T.; Klepel, S.; Renner, U.; Kostrzewa, M. Fast and reliable MALDI-TOF MS-based microorganism identification. *Nat. Methods* **2006**, *3* (4), i–ii.
77. Eigner, U.; Holfelder, M.; Oberdorfer, K.; Betz-Wild, U.; Bertsch, D.; Fahr, A.-M. Performance of a matrix-assisted laser desorption ionization-time-of-flight mass spectrometry system for the identification of bacterial isolates in the clinical routine laboratory. *Clin. Lab. (Neustadt Weinstrasse, Ger.)* **2009**, *55* (7+8), 289–296.
78. Cherkaoui, A.; Hibbs, J.; Emonet, S.; Tangomo, M.; Girard, M.; Francois, P.; Schrenzel, J. Comparison of two matrix-assisted laser desorption ionization-time of flight mass spectrometry methods with conventional phenotypic identification for routine identification of bacteria to the species level. *J. Clin. Microbiol.* **2010**, *48* (4), 1169–1175.
79. Wunschel, S. C.; Jarman, K. H.; Petersen, C. E.; Valentine, N. B.; Wahl, K. L.; Schauki, D.; Jackman, J.; Nelson, C. P.; White, E. Bacterial analysis by

MALDI-TOF mass spectrometry: An inter-laboratory comparison. *J. Am. Soc. Mass Spectrom.* **2005**, *16* (4), 456–462.

80. Karas, M.; Gluckmann, M.; Schafer, J. Ionization in matrix-assisted laser desorption/ionization: singly charged molecular ions are the lucky servivors. *J. Mass Spectrom.* **2000**, *35*, 1–12.
81. Karas, M.; Krueger, R. Ion formation in MALDI: The cluster ionization mechanism. *Chem. Rev.* **2003**, *103* (2), 427–439.
82. Beavis, R. C.; Chait, B. T. Cinnamic acid derivatives as matrices for ultraviolet laser desorption mass spectrometry of proteins. *Rapid Commun. Mass Spectrom.* **1989**, *3* (12), 432–435.
83. Fagerquist, C. K.; Garbus, B. R.; Williams, K. E.; Bates, A. H.; Harden, L. A. Covalent Attachment and Dissociative Loss of Sinapinic Acid to/from Cysteine-Containing Proteins from Bacterial Cell Lysates Analyzed by MALDI-TOF-TOF Mass Spectrometry. *J. Am. Soc. Mass Spectrom.* **2010**, *21* (5), 819–832.
84. Ilina, E. N.; Borovskaya, A. D.; Malakhova, M. M.; Vereshchagin, V. A.; Kubanova, A. A.; Kruglov, A. N.; Svistunova, T. S.; Gazarian, A. O.; Maier, T.; Kostrzewa, M.; Govorun, V. M. Direct bacterial profiling by matrix-assisted laser desorption-ionization time-of-flight mass spectrometry for identification of pathogenic Neisseria. *J. Mol. Diagn.* **2009**, *11* (1), 75–86.
85. Cohen, L.; T., C. B. Influence of matrix on the MALDI-MS analysis of peptides and proteins. *Anal. Chem.* **1996**, *68*, 31–37.
86. Kim, J.-K.; Jackson, S. N.; Murray, K. K. Matrix-assisted laser desorption/ionization mass spectrometry of collected bioaerosol particles. *Rapid Commun. Mass Spectrom.* **2005**, *19* (12), 1725–1729.
87. Ishida, Y.; Kitagawa, K.; Nakayama, A.; Ohtani, H. Complementary analysis of lipids in whole bacteria cells by thermally assisted hydrolysis and methylation-GC and MALDI-MS combined with on-probe sample pretreatment. *J. Anal. Appl. Pyrolysis* **2006**, *77* (2), 116–120.
88. Liu, H.; Du, Z.; Wang, J.; Yang, R. Universal Sample Preparation Method for Characterization of Bacteria by Matrix-Assisted Laser Desorption Ionization-Time of Flight Mass Spectrometry. *Appl. Environ. Microbiol.* **2007**, *73* (6), 1899–1907.
89. Gantt, S. L.; Valentine, N. B.; Saenz, A. J.; Kingsley, M. T.; Wahl, K. L. Use of an Internal Control for Matrix-Assisted Laser Desorption/Ionization Time-of-Flight Mass Spectrometry Analysis of Bacteria. *J. Am. Soc. Mass Spectrom.* **1999**, *10*, 1131–1137.
90. Williams, T. L.; Andrzejewski, D.; Lay, J. O.; Musser, S. M. Experimental factors affecting the quality and reproducibility of MALDI TOF mass spectra obtained from whole bacteria cells. *J. Am. Soc. Mass Spectrom.* **2003**, *14* (4), 342–51.
91. Wunschel, D. S.; Hill, E. A.; McLean, J. S.; Jarman, K.; Gorby, Y. A.; Valentine, N.; Wahl, K. Effects of varied pH, growth rate and temperature using controlled fermentation and batch culture on matrix assisted laser desorption/ionization whole cell protein fingerprints. *J. Microbiol. Methods* **2005**, *62* (3), 259–271.

92. Valentine, N.; Wunschel, S.; Wunschel, D.; Petersen, C.; Wahl, K. Effect of culture conditions on microorganism identification by matrix-assisted laser desorption ionization mass spectrometry. *Appl. Environ. Microbiol.* **2005**, *71* (1), 58–64.
93. Jarman, K. H.; Cebula, S. T.; Saenz, A. J.; Petersen, C. E.; Valentine, N. B.; Kingsley, M. T.; Wahl, K. L. An Algorithm for Automated Bacterial Identification Using Matrix-Assisted Laser Desorption/Ionization Mass Spectrometry. *Anal. Chem.* **2000**, *72* (6), 1217–1223.
94. Chen, P.; Lu, Y.; Harrington, P. B. Biomarker Profiling and Reproducibility Study of MALDI-MS Measurements of Escherichia coli by Analysis of Variance-Principal Component Analysis. *Anal. Chem.* **2008**, *80* (5), 1474–1481.
95. Horneffer, V.; Haverkamp, J.; Janssen, H.-G.; Notz, R. MALDI-TOF-MS analysis of bacterial spores: Wet heat-treatment as a new releasing technique for biomarkers and the influence of different experimental parameters and microbiological handling. *J. Am. Soc. Mass Spectrom.* **2004**, *15* (10), 1444–1454.
96. Horneffer, V.; Haverkamp, J.; Janssen, H.-G.; Ter Steeg, P. F.; Notz, R. MALDI-TOF-MS Analysis of Bacterial Spores: Wet Heat-Treatment as a New Releasing Technique for Biomarkers and the Influence of Different Experimental Parameters and Microbiological Handling. *J. Am. Soc. Mass Spectrom.* **2006**, *17* (9), 1322.
97. Magnuson, M. L.; Owens, J. H.; Kelty, C. A. Characterization of Cryptosporidium parvum by Matrix-Assisted Laser Desorption Ionization-Time of Flight Mass Spectrometry. *Appl. Environ. Microbiol.* **2000**, *66* (11), 4720–4724.
98. Amiri-Eliasi, B.; Fenselau, C. Characterization of protein biomarkers desorbed by MALDI from whole fungal cells. *Anal. Chem.* **2001**, *73* (21), 5228–5231.
99. Valentine, N. B.; Wahl, J. H.; Kingsley, M. T.; Wahl, K. L. Direct surface analysis of fungal species by matrix-assisted laser desorption/ionization mass spectrometry. *Rapid Commun. Mass Spectrom.* **2002**, *16* (14), 1352–1357.
100. Thomas, J. J.; Falk, B.; Fenselau, C.; Jackman, J.; Ezzell, J. Viral Characterization by Direct Analysis of Capsid Proteins. *Anal. Chem.* **1998**, *70* (18), 3863–3867.
101. Lasch, P.; Nattermann, H.; Erhard, M.; Staemmler, M.; Grunow, R.; Bannert, N.; Appel, B.; Naumann, D. MALDI-TOF Mass Spectrometry Compatible Inactivation Method for Highly Pathogenic Microbial Cells and Spores. *Anal. Chem.* **2008**, *80* (6), 2026–2034.
102. Lasch, P.; Beyer, W.; Nattermann, H.; Staemmler, M.; Siegbrecht, E.; Grunow, R.; Naumann, D. Identification of Bacillus anthracis by using matrix-assisted laser desorption ionization-time of flight mass spectrometry and artificial neural networks. *Appl. Environ. Microbiol.* **2009**, *75* (22), 7229–7242.
103. Seng, P.; Drancourt, M.; Gouriet, F.; La Scola, B.; Fournier, P.-E.; Rolain, J. M.; Raoult, D. Ongoing revolution in bacteriology: routine identification

of bacteria by matrix-assisted laser desorption ionization time-of-flight mass spectrometry. *Clin. Infect. Dis.* **2009**, *49* (4), 543–551.

104. Christner, M.; Rohde, H.; Wolters, M.; Sobottka, I.; Wegscheider, K.; Aepfelbacher, M. Rapid identification of bacteria from positive blood culture bottles by use of matrix-assisted laser desorption-ionization time of flight mass spectrometry fingerprinting. *J. Clin. Microbiol.* **2010**, *48* (5), 1584–1591.
105. Mellmann, A.; Bimet, F.; Bizet, C.; Borovskaya, A. D.; Drake, R. R.; Eigner, U.; Fahr, A. M.; He, Y.; Ilina, E. N.; Kostrzewa, M.; Maier, T.; Mancinelli, L.; Moussaoui, W.; Prevost, G.; Putignani, L.; Seachord, C. L.; Tang, Y. W.; Harmsen, D. High interlaboratory reproducibility of matrix-assisted laser desorption ionization-time of flight mass spectrometry-based species identification of nonfermenting bacteria. *J. Clin. Microbiol.* **2009**, *47* (11), 3732–3734.
106. Ferroni, A.; Suarez, S.; Beretti, J.-L.; Dauphin, B.; Bille, E.; Meyer, J.; Bougnoux, M.-E.; Alanio, A.; Berche, P.; Nassif, X. Real-time identification of bacteria and Candida species in positive blood culture broths by matrix-assisted laser desorption ionization-time of flight mass spectrometry. *J. Clin. Microbiol.* **2010**, *48* (5), 1542–1548.
107. van Veen, S. Q.; Claas, E. C. J.; Kuijper, E. J. High-throughput identification of bacteria and yeast by matrix-assisted laser desorption ionization-time of flight mass spectrometry in conventional medical microbiology laboratories. *J. Clin. Microbiol.* **2010**, *48* (3), 900–907.
108. Prod'hom, G.; Bizzini, A.; Durussel, C.; Bille, J.; Greub, G. Matrix-assisted laser desorption ionization-time of flight mass spectrometry for direct bacterial identification from positive blood culture pellets. *J. Clin. Microbiol.* **2010**, *48* (4), 1481–3.
109. Degand, N.; Carbonnelle, E.; Dauphin, B.; Beretti, J.-L.; Le Bourgeois, M.; Sermet-Gaudelus, I.; Segonds, C.; Berche, P.; Nassif, X.; Ferroni, A. Matrix-assisted laser desorption ionization-time of flight mass spectrometry for identification of nonfermenting gram-negative bacilli isolated from cystic fibrosis patients. *J. Clin. Microbiol.* **2008**, *46* (10), 3361–3367.
110. Bizzini, A.; Durussel, C.; Bille, J.; Greub, G.; Prod'hom, G. Performance of matrix-assisted laser desorption ionization-time of flight mass spectrometry for identification of bacterial strains routinely isolated in a clinical microbiology laboratory. *J. Clin. Microbiol.* **2010**, *48* (5), 1549–1554.
111. Vanlaere, E.; Sergeant, K.; Dawyndt, P.; Kallow, W.; Erhard, M.; Sutton, H.; Dare, D.; Devreese, B.; Samyn, B.; Vandamme, P. Matrix-assisted laser desorption ionization-time-of-flight mass spectrometry of intact cells allows rapid identification of Burkholderia cepacia complex. *J. Microbiol. Methods* **2008**, *75* (2), 279–286.
112. Alispahic, M.; Hummel, K.; Jandreski-Cvetkovic, D.; Nobauer, K.; Razzazi-Fazeli, E.; Hess, M.; Hess, C. Species-specific identification and differentiation of Arcobacter, Helicobacter and Campylobacter by full-spectral matrix-associated laser desorption/ionization time of flight mass spectrometry analysis. *J. Med. Microbiol.* **2010**, *59* (3), 295–301.

113. Moliner, C.; Ginevra, C.; Jarraud, S.; Flaudrops, C.; Bedotto, M.; Couderc, C.; Etienne, J.; Fournier, P.-E. Rapid identification of Legionella species by mass spectrometry. *J. Med. Microbiol.* **2010**, *59* (3), 273–284.
114. Fournier, P.-E.; Couderc, C.; Buffet, S.; Flaudrops, C.; Raoult, D. Rapid and cost-effective identification of Bartonella species using mass spectrometry. *J. Med. Microbiol.* **2009**, *58* (9), 1154–1159.
115. Madonna, A. J.; Basile, F.; Furlong, E.; Voorhees, K. J. Detection of bacteria from biological mixtures using immunomagnetic separation combined with matrix-assisted laser desorption/ionization time-of-flight mass spectrometry. *Rapid Commun. Mass Spectrom.* **2001**, *15* (13), 1068–74.
116. Madonna, A. J.; Van Cuyk, S.; Voorhees, K. J. Detection of Escherichia coli using immunomagnetic separation and bacteriophage amplification coupled with matrix-assisted laser desorption/ionization time-of-flight mass spectrometry. *Rapid Commun. Mass Spectrom.* **2003**, *17* (3), 257–63.
117. Bundy, J.; Fenselau, C. Lectin-Based Affinity Capture for MALDI-MS Analysis of Bacteria. *Anal. Chem.* **1999**, *71*, 1460–1463.
118. Bundy, J. L.; Fenselau, C. Lectin and carbohydrate affinity capture surfaces for mass spectrometric analysis of microorganisms. *Anal. Chem.* **2001**, *73* (4), 751–7.
119. Afonso, C.; Fenselau, C. Use of bioactive glass slides for matrix-assisted laser desorption/ionization analysis: application to microorganisms. *Anal. Chem.* **2003**, *75* (3), 694–7.
120. Whiteaker, J.; Karns, J.; Fenselau, C.; Perdue, M. L. Analysis of Bacillus anthracis spores in milk using mass spectrometry. *Foodborne Pathog. Dis.* **2004**, *1* (3), 185–194.
121. Hazen, T. H.; Martinez, R. J.; Chen, Y.; Lafon, P. C.; Garrett, N. M.; Parsons, M. B.; Bopp, C. A.; Sullards, M. C.; Sobecky, P. A. Rapid identification of Vibrio parahaemolyticus by whole-cell matrix-assisted laser desorption ionization-time of flight mass spectrometry. *Appl. Environ. Microbiol.* **2009**, *75* (21), 6745–6756.
122. Barbuddhe, S. B.; Maier, T.; Schwarz, G.; Kostrzewa, M.; Hof, H.; Domann, E.; Chakraborty, T.; Hain, T. Rapid identification and typing of Listeria species by matrix-assisted laser desorption ionization-time of flight mass spectrometry. *Appl. Environ. Microbiol.* **2008**, *74* (17), 5402–5407.
123. Fernandez-No, I. C.; Boehme, K.; Gallardo, J. M.; Barros-Velazquez, J.; Canas, B.; Calo-Mata, P. Differential characterization of biogenic amine-producing bacteria involved in food poisoning using MALDI-TOF mass fingerprinting. *Electrophoresis* **2010**, *31* (6), 1116–1127.
124. Parisi, D.; Magliulo, M.; Nanni, P.; Casale, M.; Forina, M.; Roda, A. Analysis and classification of bacteria by matrix-assisted laser desorption/ionization time-of-flight mass spectrometry and a chemometric approach. *Anal. Bioanal. Chem.* **2008**, *391* (6), 2127–2134.
125. MacFadden, J. F. *Biochemical tests for identification of medical bacteria*, 3rd ed.; Lippincott Williams & Wilkins: 2000; p 912.
126. Bohme, K.; Fernandez-No, I. C.; Barros-Velazquez, J.; Gallardo, J. M.; Calo-Mata, P.; Canas, B. Species Differentiation of Seafood Spoilage and

Pathogenic Gram-Negative Bacteria by MALDI-TOF Mass Fingerprinting. *J. Proteome Res.* **2010**, *9* (6), 3169–3183.
127. Salih, E. Phosphoproteomics by mass spectrometry and classical protein chemistry approaches. *Mass Spectrom. Rev.* **2005**, *24* (6), 828–846.
128. Ho, K.-C.; Tsai, P.-J.; Lin, Y.-S.; Chen, Y.-C. Using Biofunctionalized Nanoparticles To Probe Pathogenic Bacteria. *Anal. Chem.* **2004**, *76* (24), 7162–7168.
129. Lin, Y.-S.; Tsai, P.-J.; Weng, M.-F.; Chen, Y.-C. Affinity Capture Using Vancomycin-Bound Magnetic Nanoparticles for the MALDI-MS Analysis of Bacteria. *Anal. Chem.* **2005**, *77* (6), 1753–1760.
130. Liu, J.-C.; Tsai, P.-J.; Lee, Y. C.; Chen, Y.-C. Affinity Capture of Uropathogenic Escherichia coli Using Pigeon Ovalbumin-Bound Fe3O4@Al2O3 Magnetic Nanoparticles. *Anal. Chem.* **2008**, *80* (14), 5425–5432.
131. Guo, Z.; Liu, Y.; Li, S.; Yang, Z. Interaction of bacteria and ion-exchange particles and its potential in separation for matrix-assisted laser desorption/ionization mass spectrometric identification of bacteria in water. *Rapid Commun. Mass Spectrom.* **2009**, *23* (24), 3983–3993.
132. Li, S.; Guo, Z.; Liu, Y.; Yang, Z.; Hui, H. K. Integration of microfiltration and anion-exchange nanoparticles-based magnetic separation with MALDI mass spectrometry for bacterial analysis. *Talanta* **2009**, *80* (1), 313–320.
133. Li, S.; Guo, Z.; Wu, H.-F.; Liu, Y.; Yang, Z.; Woo, C. H. Rapid analysis of Gram-positive bacteria in water via membrane filtration coupled with nanoprobe-based MALDI-MS. *Anal. Bioanal. Chem.* **2010**, *397* (6), 2465–2476.
134. Dworzanski, J. P.; Snyder, A. P. Classification and identification of bacteria using mass spectrometry-based proteomics. *Expert Rev. Proteom.* **2005**, *2* (6), 863–878.
135. Aebersold, R.; Goodlett, D. R. Mass Spectrometry in Proteomics. *Chem. Rev.* **2000**.
136. Aebersold, R.; Mann, M. Mass Spectrometry-Based Proteomics. *Nature* **2003**, *422* (13 March), 198–208.
137. Yao, Z.-P.; Demirev, P. A.; Fenselau, C. Mass spectrometry-based proteolytic mapping for rapid virus identification. *Anal. Chem.* **2002**, *74* (11), 2529–2534.
138. Lill, J. R.; Ingle, E. S.; Liu, P. S.; Pham, V.; Sandoval, W. N. Microwave-assisted proteomics. *Mass Spectrom. Rev.* **2007**, *26* (5), 657–71.
139. Hauser, N. J.; Basile, F. Development of a Rapid Non-Enzymatic Cell Digestion Procedure for Proteomics Based Microorganism Identification: Site-Specific Digestion at Aspartyl Residue by Microwave Heating - Mild Acid Hydrolysis. Presented at 52nd ASMS Meeting, Nashville, TN, 2004.
140. Swatkoski, S.; Russell, S. C.; Edwards, N.; Fenselau, C. Rapid Chemical Digestion of Small Acid-Soluble Spore Proteins for Analysis of Bacillus Spores. *Anal. Chem.* **2006**, *78* (1), 181–188.
141. Yao, Z.-P.; Afonso, C.; Fenselau, C. Rapid microorganism identification with on-slide proteolytic digestion followed by matrix-assisted laser

desorption/ionization tandem mass spectrometry and database searching. *Rapid Commun. Mass Spectrom.* **2002**, *16* (20), 1953–1956.
142. Warscheid, B.; Fenselau, C. A targeted proteomics approach to the rapid identification of bacterial cell mixtures by matrix-assisted laser desorption/ionization mass spectrometry. *Proteomics* **2004**, *4* (10), 2877–2892.
143. Dugas, A. J., Jr.; Murray, K. K. On-target digestion of collected bacteria for MALDI mass spectrometry. *Anal. Chim. Acta* **2008**, *627* (1), 154–161.
144. Chen, W.-J.; Tsai, P.-J.; Chen, Y.-C. Functional Nanoparticle-Based Proteomic Strategies for Characterization of Pathogenic Bacteria. *Anal. Chem.* **2008**, *80* (24), 9612–9621.
145. Cech, N. B.; Enke, C. G. Practical implications of some recent studies in electrospray ionization fundamentals. *Mass Spectrom. Rev.* **2001**, *20* (6), 362–87.
146. Dworzanski, J. P.; Snyder, A. P.; Chen, R.; Zhang, H.; Wishart, D.; Li, L. Identification of bacteria using tandem mass spectrometry combined with a proteome database and statistical scoring. *Anal. Chem.* **2004**, *76* (8), 2355–2366.
147. Dworzanski, J. P.; Dickinson, D. N.; Deshpande, S. V.; Snyder, A. P.; Eckenrode, B. A. Discrimination and Phylogenomic Classification of Bacillus anthracis-cereus-thuringiensis Strains Based on LC-MS/MS Analysis of Whole Cell Protein Digests. *Anal. Chem.* **2010**, *82* (1), 145–155.
148. Everley, R. A.; Mott, T. M.; Wyatt, S. A.; Toney, D. M.; Croley, T. R. Liquid Chromatography/Mass Spectrometry Characterization of Escherichia coli and Shigella Species. *J. Am. Soc. Mass Spectrom.* **2008**, *19* (11), 1621–1628.
149. Mott, T. M.; Everley, R. A.; Wyatt, S. A.; Toney, D. M.; Croley, T. R. Comparison of MALDI-TOF/MS and LC-QTOF/MS methods for the identification of enteric bacteria. *Int. J. Mass Spectrom.* **2010**, *291* (1–2), 24–32.
150. Fontana, A.; Dalzoppo, D.; Grandi, C.; Zambonin, M. Cleavage at tryptophan with o-iodosobenzoic acid. *Methods Enzymol.* **1983**, *91* (Enzyme Struct., Pt. I), 311–18.
151. Huang, H. V.; Bond, M. W.; Hunkapiller, M. W.; Hood, L. E. Cleavage at tryptophanyl residues with dimethyl sulfoxide-hydrochloric acid and cyanogen bromide. *Methods Enzymol.* **1983**, *91* (Enzyme Struct., Pt. I), 318–24.
152. Inglis, A. S. Cleavage at aspartic acid. *Methods in Enzymology* **1983**, *91* (Enzyme Struct., Part I), 324–32.
153. Li, A.; Sowder, R. C., II; Henderson, L. E.; Moore, S. P.; Garfinkel, D. J.; Fisher, R. J. Chemical cleavage at aspartyl residues for protein identification. *Anal. Chem.* **2001**, *73* (22), 5395–5402.
154. Fenselau, C.; Laine, O.; Swatkoski, S. Microwave assisted acid cleavage for denaturation and proteolysis of intact human adenovirus. *Int. J. Mass Spectrom.* **2010**, in press, corrected proof.
155. Hauser, N. J.; Basile, F. On-line Microwave D-Cleavage LC-ESI-MS/MS of Intact Proteins: Site-Specific Cleavages at Aspartic Acid Residues and Disulfide Bonds. *J. Proteome Res.* **2008**, *7*, 1012–1026.

156. Wells, J. M.; McLuckey, S. A. Collision-induced dissociation (CID) of peptides and proteins. *Methods in Enzymology* **2005**, *402* (Biological Mass Spectrometry), 148–185.
157. Stokes, A. A.; Clarke, D. J.; Weidt, S.; Langridge-Smith, P.; MacKay, C. L. Top-down protein sequencing by CID and ECD using desorption electrospray ionization (DESI) and high-field FTICR mass spectrometry. *Int. J. Mass Spectrom.* **2009**, *289* (1), 54–57.
158. Demirev, P. A.; Feldman, A. B.; Kowalski, P.; Lin, J. S. Top-Down Proteomics for Rapid Identification of Intact Microorganisms. *Anal. Chem.* **2005**, *77* (22), 7455–7461.
159. Zubarev, R. A.; Horn, D. M.; Fridriksson, E. K.; Kelleher, N. L.; Kruger, N. A.; Lewis, M. A.; Carpenter, B. K.; McLafferty, F. W. Electron Capture Dissociation for Structural Characterization of Multiply Charged Protein Cations. *Anal. Chem.* **2000**, *72* (3), 563–573.
160. Ge, Y.; Lawhorn, B. G.; ElNaggar, M.; Strauss, E.; Park, J. H.; Begley, T. P.; McLafferty, F. W. Top down characterization of larger proteins (45 kDa) by electron capture dissociation mass spectrometry. *J. Am. Chem. Soc.* **2002**, *124* (4), 672–8.
161. Mikesh, L. M.; Ueberheide, B.; Chi, A.; Coon, J. J.; Syka, J. E. P.; Shabanowitz, J.; Hunt, D. F. The utility of ETD mass spectrometry in proteomic analysis. *Biochim. Biophys. Acta, Proteins Proteomics* **2006**, *1764* (12), 1811–1822.
162. Bunger, M. K.; Cargile, B. J.; Ngunjiri, A.; Bundy, J. L.; Stephenson, J. L., Jr. Automated Proteomics of E. coli via Top-Down Electron-Transfer Dissociation Mass Spectrometry. *Anal. Chem.* **2008**, *80* (5), 1459–1467.
163. Wynne, C.; Fenselau, C.; Demirev, P. A.; Edwards, N. Top-Down Identification of Protein Biomarkers in Bacteria with Unsequenced Genomes. *Anal. Chem.* **2009**, *81* (23), 9633–9642.
164. Wynne, C.; Edwards, N. J.; Fenselau, C. Phyloproteomic classification of unsequenced organisms by top-down identification of bacterial proteins using capLC-MS/MS on an Orbitrap. *Proteomics* **2010**, *10* (20), 3631–3643.
165. Warscheid, B.; Fenselau, C. Characterization of Bacillus spore species and their mixtures using postsource decay with a curved-field reflectron. *Anal. Chem.* **2003**, *75* (20), 5618–5627.
166. Fagerquist, C. K.; Garbus, B. R.; Miller, W. G.; Williams, K. E.; Yee, E.; Bates, A. H.; Boyle, S.; Harden, L. A.; Cooley, M. B.; Mandrell, R. E. Rapid Identification of Protein Biomarkers of Escherichia coli O157:H7 by Matrix-Assisted Laser Desorption Ionization-Time-of-Flight-Time-of-Flight Mass Spectrometry and Top-Down Proteomics. *Anal. Chem.* **2010**, *82* (7), 2717–2725.
167. McLuckey, S. A. The emerging role of ion/ion reactions in biological mass spectrometry: considerations for reagent ion selection. *Eur. J. Mass Spectrom.* **2010**, *16* (3), 429–436.
168. Mukhopadhyay, R. MS makes identification of clinical pathogens a cinch. *Anal. Chem.* **2009**, *81* (23), 9537.

Chapter 3

Discrimination of Fungi by MALDI-TOF Mass Spectrometry

Justin M. Hettick,* Brett J. Green, Amanda D. Buskirk, James E. Slaven, Michael L. Kashon, and Donald H. Beezhold

Centers for Disease Control and Prevention, National Institute for Occupational Safety and Health, Health Effects Laboratory Division, Morgantown, WV
***jhettick@cdc.gov**

Traditionally, fungal identification has largely been based on the subjective micro- and macroscopic examination of morphological and culture characteristics. Matrix-assisted laser desorption/ionization time-of-flight mass spectrometry (MALDI TOF MS) was used to generate reproducible mass spectral "fingerprints" for 76 fungal species, particularly from the medically important genera, *Penicillium* and *Aspergillus*. The mass spectra contain abundant mass signals and allow unambiguous discrimination between species. Species identification error rates were determined to be 0% and 1.4% using resubstitution and cross-validation methods, respectively. The ability of MALDI TOF MS to differentiate fungal strains was additionally examined for the aflatoxin producing species, *Aspergillus flavus*. Identification error rates for 40 tested *A. flavus* cultures from five unique strains were determined to be 0% and 5% using resubstitution and cross-validation methods, respectively. Analysis of dematiaceous (dark-pigmented) fungi has been observed to yield poor MALDI-TOF mass spectra. Results demonstrate this was due to the presence of melanin in the cell wall of fungal spores and hyphae. Strategies to overcome this limitation are presented. These results indicate that MALDI-TOF MS data may be a useful diagnostic tool and alternative to available immunodiagnostic and molecular methods for the objective identification of environmental, industrial and clinically important fungal species.

Introduction

Fungi are a diverse group of heterotrophic eukaryotes that disseminate a variety of bioaerosols into the environment. Fungal spores, hyphae, and fragments can be ubiquitous in indoor, outdoor, and occupational environments (*1–3*) and are among the most common bioaerosols that humans inhale (*2*). Fungal bioaerosols contain proteins, secondary metabolites, mycotoxins, β(1,3)-D-glucan, chitin, and volatile organic compounds (*2*, *4*, *5*) that may be a burden to public health, particularly in indoor and occupational contaminated environments (*1*, *5*). Personal exposure has been associated with exacerbations of adverse health effects ranging from allergic rhinitis, asthma, hypersensitivity pneumonitis, dermatitis, invasive aspergillosis, to death (*5–10*). Some dimorphic fungi may even act as invasive pathogens in patients that are immunocompetent (*11*). Viable and nonviable methodologies have been traditionally used to detect and quantify fungi associated with adverse health effects; however, many of these techniques have been confounded by bias and subjectivity (*2*). Due to the health and economic impacts associated with fungal exposure, a number of new molecular and proteomic technologies have been developed that have improved the identification and characterization of medically important fungi.

Traditional viable and non-viable methods of fungal identification rely on the subjective identification of micro- and macroscopic morphological culture and spore characteristics. As such, these evaluations rely on the taxonomic judgment of a trained mycologist and are therefore subject to observer bias. A number of studies have shown that morphological similarities exist between numerous genera, species and strains, and as a result, fungal misclassifications are a confounding variable associated with viable and nonviable analyses (*12–14*). Emerging immunodiagnostic and molecular technologies for rapid identification of microorganisms, including mass spectrometry-based techniques, particularly those based on matrix-assisted laser desorption/ionization (MALDI) (*15*, *16*) time-of-flight mass spectrometry (TOF MS) (*17*) are particularly promising. Mass spectrometry-based techniques have provided a rapid (MS analysis takes minutes) and sensitive alternative method for identifying as few as 10^3 fungal cells.

A number of studies have been presented in the literature on the analysis of intact cells by MALDI-TOF MS. Although significantly more attention has been paid to bacteria (*18–29*), recent studies have focused on the identification of medically important fungi (*30–36*). Of particular interest for rapid identification are so called "fingerprint" methods (*24*, *37*, *38*) that utilize pattern recognition to correlate an unknown mass spectrum with a known organism from a library of spectra.

Early experiments in our laboratory focused on utilizing MALDI-TOF MS data coupled with biostatistical analysis to discriminate between *Mycobacterium* species (*28*). Discrimination was possible at both the species and strain level (*29*). The results of our research and that of other groups (*26*) suggested that care must be taken to carefully define both the culture conditions and mass spectrometry parameters in order to achieve highly reproducible mass spectrometry fingerprints from microorganisms. Any changes to the experimental conditions may alter the appearance of the MALDI-TOF mass spectrum and affect the results of the identification. Among the variables to be considered in MALDI TOF analysis are culture time, nutrient media, microorganism concentration, and cell lysis. Similarly, mass spectrometry parameters such as choice of matrix, desorption laser fluence, and delayed extraction settings must be maintained. Consistency in experimental parameters is the key to reproducible fingerprint mass spectra.

Furthermore, because a biological specimen of interest may be a pathogen or aeroallergen, biosafety is of paramount concern. For this reason, our laboratory avoids "whole cell" methods where the microorganism of interest is deposited directly on the sample stage and introduced to the mass spectrometer. Rather, microorganisms are handled and extracted in a biological safety cabinet and extract preparations are subsequently analyzed. Early experiments in our laboratory demonstrated that the data acquired from extracts that mimic the solvent composition of the MALDI matrix solution produce high-quality results, similar to "whole cell" spectra (*28*). In addition, our previous experiments (*35*) demonstrated that for fungal analysis, a bead-beating cellular disruption step during the chemical extraction increases both the number of peaks and the signal-to-noise ratio observed in MALDI-TOF fingerprint mass spectra. Similar approaches to cell lysis, such as exposure to acid, ultrasonication, and corona plasma discharge, have been previously applied by other researchers with success (*30*).

In this study, we have applied MALDI-TOF MS fingerprinting methods to an extensive library of fungal isolates covering more than sixty-eight species in thirty-five genera to evaluate our methodology for fungal discrimination using a library covering a broad range of genera. Furthermore, we have previously observed that certain dematiaceous (dark-pigmented) fungi yield poor MALDI-TOF fingerprint mass spectra (*34*). Here we present data supporting the hypothesis that fungal melanins, which have both photo- and chemoprotective properties in fungi, have a suppressive effect on MALDI-TOF fingerprint mass spectra.

Experimental

Reagents

Angiotensin II (human), insulin oxidized B chain (bovine), cytochrome C (equine), albumin (bovine serum), α-cyano-4-hydroxycinnamic acid (CHCA) and trifluoroacetic acid (TFA) were purchased from Sigma-Aldrich (St. Louis, MO). Acetonitrile (HPLC grade) was purchased from Fisher Scientific (Fairlawn, NJ). Malt extract agar was purchased from Difco (Sparks, MD). Tricyclazole (5-methyl-1,2,4-triazolo[3,4-*b*][1,3] benzothiazole) was purchased from Wako Pure Chemical Industries (Osaka, Japan). Deionized water was produced by a Millipore Synthesis A-10 (Billerica, MA).

Fungal Culture

Fungal Fingerprinting

Seventy-six fungal isolates from a variety of genera (Table I) were sub-cultured from NIOSH or American Type Culture Collection (ATCC, Manassas, VA) stock sources and grown for 14 days at 25 °C on malt extract agar (MEA). To ensure reproducibility, eight independent cultures were performed for each isolate, for a total of 608 individual fungal cultures. Conidia and hyphae from one culture plate (~10^8 cells) were transferred to 100 μL of 0.1 mm zirconium beads (Biospec, Bartlesville, OK) and 1 mL 50/50 acetonitrile/4% trifluoroacetic acid. The samples were subjected to three one-minute bead-beating cycles. The resulting solutions were centrifuged at 8,800 x g for 10 minutes and the supernatant taken for MALDI-TOF MS analysis.

Table I. List of fungal isolates cultured for MALDI analysis

Genus	*No. of Species/Strains*	*No. of cultures*
Acremonium	1	8
Alternaria	2	16
Aspergillus	18	144
Aureobasidium	1	8
Candida	1	8
Chaetomium	1	8
Cladosporium	3	24
Cochliobolus	1	8
Cryptococcus	1	8
Curvularia	1	8
Emericella	1	8
Epicoccum	1	8
Eurotium	2	16
Exserohilum	1	8
Fusarium	3	24
Geotrichum	1	8
Hansenula	1	8
Memnoniella	1	8
Mucor	1	8
Myrothecium	1	8

Continued on next page.

Table I. (Continued). List of fungal isolates cultured for MALDI analysis

Genus	*No. of Species/Strains*	*No. of cultures*
Neosartorya	1	8
Paecilomyces	3	24
Penicillium	12	96
Pithomyces	1	8
Phoma	1	8
Rhizopus	2	16
Rhodotorula	1	8
Saccharomyces	1	8
Scopulariopsis	1	8
Stachybotrys	3	24
Stemphylium	1	8
Talaromyces	1	8
Trichoderma	2	16
Ulocladium	2	16
Wallemia	1	8
TOTAL	**76**	**608**

Suppression of Fungal Melanin Production in Culture

Eight replicate cultures of *Aspergillus niger* were grown for 7 days at 25 °C on malt extract agar and malt extract agar supplemented with 1% ethanol containing 50 μg/mL tricyclazole. Conidia and hyphae from one culture plate (~10^8 cells) were transferred to 100 μL of 0.1 mm zirconium beads (Biospec, Bartlesville, OK) and 1 mL of 50/50 acetonitrile/4% trifluoroacetic acid. Samples were subjected to three one-minute bead-beating cycles. The resulting solutions were centrifuged at 8,800 x g for 10 minutes and the supernatant taken for MALDI-TOF MS analysis.

Preparation of Melanin "Ghosts"

Melanin "ghosts" were produced from *Aspergillus niger* according to the method published by Wang, et al. (*39*). In brief, *A. niger* was grown on malt extract agar for ten days at 25 °C. Sporulating fungal cultures were harvested from the culture plate using sterilized deionized water (DI). The resulting suspension was centrifuged at 1000 x g for ten minutes, and the supernatant discarded. The pellet was resuspended in sodium citrate buffer containing 10 mg/mL of cell lysing enzymes from *T. harzianum* and 2 mg/mL cellulase from *T. reesei* and incubated overnight with agitation at 30 °C. The sample was then centrifuged at

1000 x g for ten minutes and the supernatant discarded. The sample was washed with phosphate buffered saline (PBS) and centrifuged for ten minutes at 1000 x g. The pellet was resuspended in 4 M guanidine isothiocyanate and incubated at room temperature over night with agitation. The sample was washed with PBS as described above and the pellet resuspended in 1 mg/mL of proteinase K in DI and incubated overnight at 37°C with agitation. The sample was washed with PBS and the pellet resuspended in 6 M HCl and boiled for one hour. The sample was then washed with PBS a final time and resuspended in PBS. The resulting final melanin extract was dialyzed against DI H_2O for ten days at 4°C using 3500 Da molecular weight cutoff (MWCO) membrane tubing (Spectra/Por®, Laguna Hills, CA, USA). The sample was then lyophilized and the resulting powder was used for further experiments.

The *A. niger* melanin ghosts were diluted with 50/50 ACN/DI and mixed with an equal volume of 0.5 mg/mL HSA. The final concentration of melanin ghosts was approximately 0.1 mg/mL. The resulting solution was mixed with an equal volume of 10 mg/mL α-cyano-4-hydroxycinnamic acid and 1 μL aliquots deposited on a gold sample stage (Bio-Rad, Hercules, CA, USA) and allowed to air dry prior to MALDI analysis.

Mass Spectrometry

MALDI-TOF MS samples were prepared by mixing supernatant 1:1 with 10 mg/mL α-cyano-4-hydroxycinnamic acid in 50/50 acetonitrile/0.1% trifluoroacetic acid. 1 μL of the resulting solution was deposited on a gold sample stage (Bio-Rad, Hercules, CA) and allowed to air dry. Each sample was analyzed in duplicate, for a total of sixteen composite MALDI-TOF mass spectra per isolate. MALDI-TOF mass spectra were acquired using a Ciphergen PBS-IIc linear time-of-flight mass spectrometer (Bio-Rad, Hercules, CA) with a flight path of 0.8 m, capable of mass resolution ($m/\Delta m$) of 500-1000 and a mass accuracy of ± 1000 ppm. Spectra were acquired over the *m/z* range 0-100 kDa, with the delayed extraction parameters set to optimally focus the 10-20 kDa range. Composite mass spectra are the average of 100 laser shots taken from 20 distinct positions across the sample deposit. These positions were held constant for all samples. N_2 laser (337-nm) intensity was maintained just above the threshold for ion production (laser step 140-160). Mass spectra were externally calibrated using a set of peptide and protein calibrants that covered the range of 1-66 kDa.

Data Analysis

Initial data analysis was performed using the Biomarker Wizard (Bio-Rad, Hercules, CA) software suite. Spectra were baseline corrected and normalized to total ion current. "Clusters" of peaks common to a given isolate were generated by selecting all peaks with signal-to-noise (S/N) greater than 5 that occurred in each spectrum from that isolate. The mass tolerance for each cluster was set to 0.3% of the m/z. Linear discriminant functions were analyzed using SAS/STAT software, Version 9.1 of the SAS system for Windows (SAS Institute, Cary, NC). The intensity values were first tested for their distribution, and were found to be log-normal, so a natural log transformation was utilized on the intensity values. A stepwise variable selection method using the "PROC STEPDISC" procedure, which selects a subset of the variables of interest using a stepwise discriminant analysis, keeping the most significant variables from iterative F-tests, was then performed to select a subset of variables that could serve as predictor variables for class membership. Using this new subset of significant peaks, "PROC DISCRIM" was utilized to determine the classification error rate. This was done by calculating a discriminant function that allowed each data point to be compared to all others for cross-validation classification. This process was performed iteratively for each data point individually. This procedure was followed by the "PROC CANDISC" procedure, a procedure that reduces the number of dimensions to find linear combinations of the variable set that also summarizes between-class variation, to perform a canonical discriminant analysis on that subset of variables. This creates new variables by taking linear combinations of the original variables and aids in determining the true underlying dimension of the data space. The canonical functions generated allow the calculation of canonical scores, which can be used to discriminate among the various isolates. Using the same subsets of variables, cluster analysis was then performed using the "PROC CLUSTER" procedure, which uses distances between data points to form hierarchical clusters, and a dendrogram was generated using the "PROC TREE" procedure, a procedure that uses the data set from "PROC CLUSTER" to produce a dendrogram.

Results and Discussion

Fungal Fingerprinting

Seventy-six fungal isolates from thirty-five genera were independently cultured in order to perform MALDI-TOF MS fingerprinting (Table I). Included in this library were twelve species derived from the genera *Penicillium* and *Aspergillus*, as well as five unique strains of the aflatoxin producing species, *A. flavus*. Six hundred and eight independent cultures were analyzed in duplicate by MALDI-TOF MS, resulting in a database of 1216 individual mass spectra. Representative spectra from five species are presented in Figure 1. In general, fingerprint mass spectra from fungi were characterized by several abundant signals in the region of 5-25 kDa, however, several fungi exhibited signals as high in *m/z* as 40-45 kDa. Many fungi exhibited a strong peak at approximately 8.5 kDa, a peak which has been tentatively identified as ubiquitin, an abundant protein that has been suggested to be a biomarker for the eukaryotic kingdom (*30*).

Initial statistical analysis of the complete database of fungal fingerprint mass spectra identified 1422 peaks of S/N greater than 5. The "STEPDISC" stepwise variable selection method within the SAS/STAT software identified a subset of 181 significant peaks to utilize for discrimination within the dataset. Both resubstitution and cross-validation methods were used to test how well this subset of 181 peaks could discriminate between the 68 species present in the database. The resubstitution method resulted in a 0% error rate, whereas the cross-validation method resulted in a 1.47% error rate. In resubstitution, the discriminant function is fitted to the dataset and then applied to each observation.

In contrast, cross-validation deletes the observation, fits the discriminant function to the remaining dataset, and then applies the function to the deleted observation. Resubstitution tends to underestimate classification error, whereas cross-validation is unbiased and preferred for large datasets (*40*). The ability to match an acquired fingerprint mass spectrum to one of 68 species in the database with 0-1.5% error underscores the utility of the methodology for fungal identification.

In order to further illustrate the utility of this methodology for species discrimination, several subsets of the full database were queried in more detail. If just the subset of the database corresponding to the species within the genus *Aspergillus* is examined, discrimination on the basis of the raw statistical data (S/N > 5) provides an error rate of 0 or 18% (resubstitution and cross-validation, respectively), however, when just the significant peaks are utilized, both methods return 0% error rates for classification. The discriminatory power of this method can be illustrated by plotting canonical variables for this subset of the database. Figure 2 presents a three-dimensional plot of the first three canonical variables for each species of *Aspergillus*.

The three canonical variables for each independent culture shown in Figure 2 are plotted individually and represented by a circle. There are eight observations for each species, one for each individual culture. In each case, the eight observations cluster tightly together in three-dimensional space and are spatially resolved from the other *Aspergillus* species. These data demonstrate that the mass-abundance data derived from the MALDI-TOF MS fingerprint spectrum of each independent culture are highly reproducible and distinct from other similar species of the same genera.

Similarly, the error rates for species classification were calculated on the subset of the database comprised of twelve species belonging to the genus, *Penicillium*. In this case, both resubstitution and cross-validation methods returned 0% error rates for species identification within the *Penicillium* subset using both the raw MALDI-TOF MS data (S/N > 5) and the stepwise-selected significant variables. The three-dimensional canonical discriminant data for the twelve species of *Penicillium* are presented in Figure 3. Similar to the *Aspergillus* data presented in Figure 2, the eight independent cultures of each species of *Penicillium* cluster tightly together and are distinct from other species within the genus. It should be noted that although a few clusters in Figure 3 appear not to be resolved, this figure utilizes only three canonical variables for ease of visualization. More canonical variables exist and can be used to unambiguously discriminate these species.

In addition to discrimination of species within a genus, the MALDI-TOF MS fingerprint method may be used to discriminate between strains of the same species. Five strains of *Aspergillus flavus* were independently cultured and analyzed by MALDI-TOF fingerprint analysis. Although the MALDI-TOF fingerprint mass spectra for the five species were very similar in terms of observed *m/z* (*34*), reproducible variations in the relative abundance of observed *m/z* ratios allows unambiguous discrimination. Figure 4 presents the three-dimensional canonical discriminant data for the *A. flavus* subset of the database. It is important to note that this discrimination is possible based solely on the differential

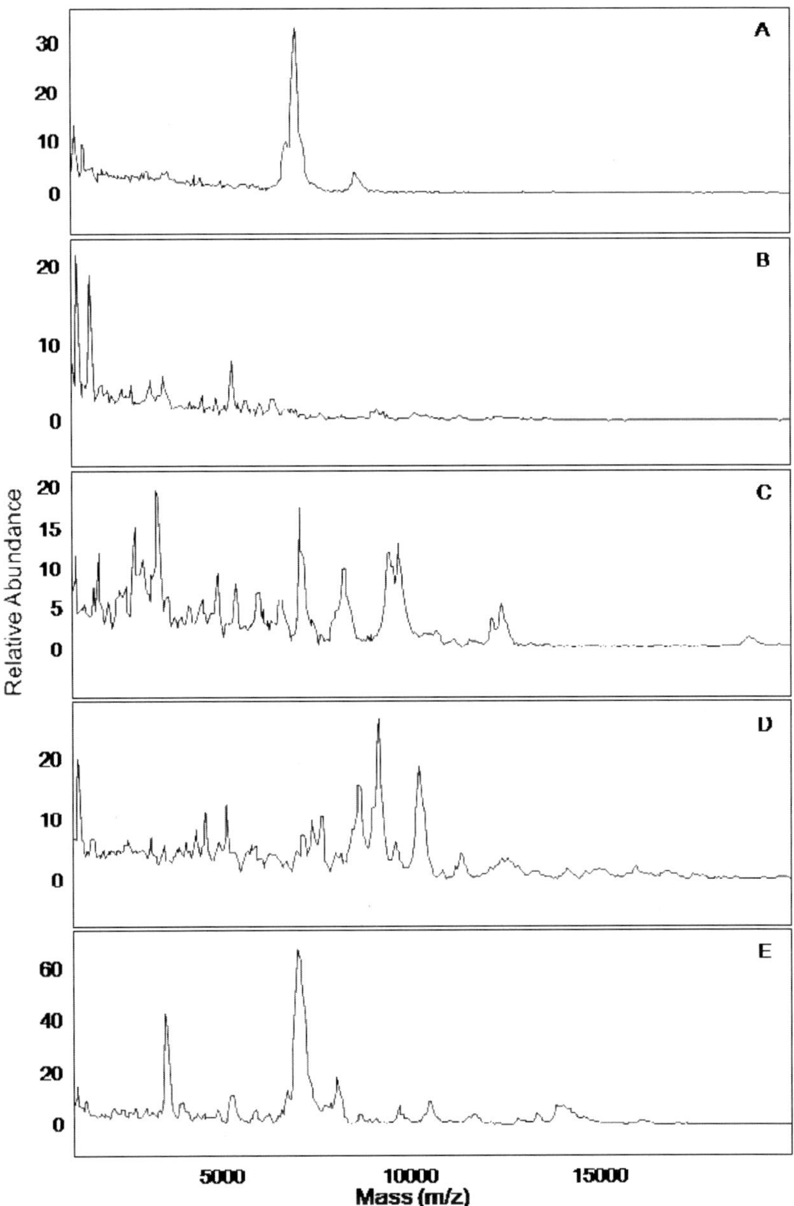

Figure 1. MALDI-TOF MS fingerprint spectra of selected fungi from the dataset. (A) C. herbarum; (B) A. alternata; (C) P. variotii; (D) A. brassicola; (E) C. albicans.

expression of shared *m/z*, rather than a difference in observed m/z as is the case when differentiating between species/genera. Resubstitution and cross-validation methods produce error rates for strain identification of 0 and 5% respectively.

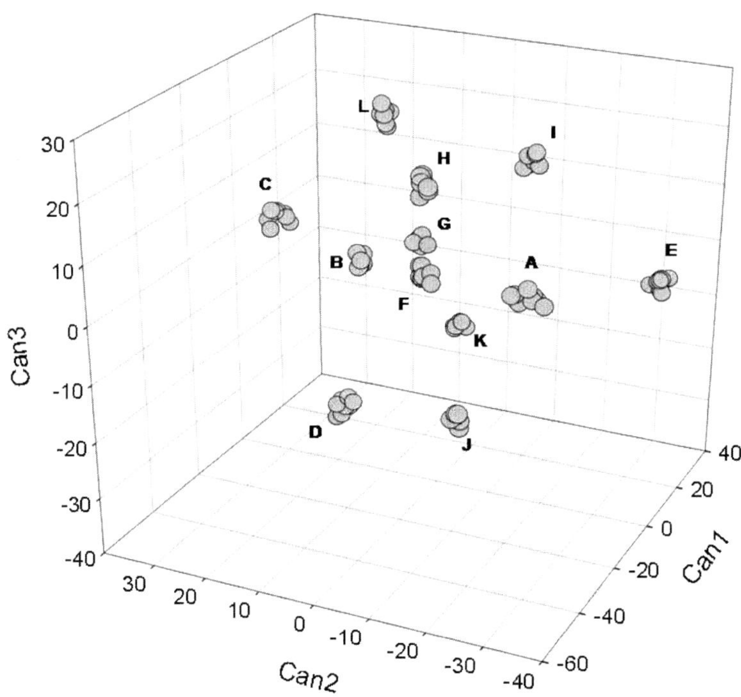

Figure 2. Three-dimensional canonical discriminant plot for MALDI-TOF MS data derived from Aspergillus species from the whole fungal dataset. (A) A. candidus; (B) A. chevalieri; (C) A. flavus; (D) A. fumigatus; (E) A. nidulans; (F) A. niger; (G) A. parasiticus; (H) A. repens; (I) A. sydowii; (J) A. terreus; (K) A. ustus; (L) A. versicolor.

Suppression of MALDI-TOF MS Signal by Fungal Melanin

While constructing the library of MALDI-TOF MS fingerprint mass spectra from 76 fungal isolates, it was noted that certain species yielded very poor fingerprint spectra in which very few or no peaks were observed. Examination of the database demonstrated that this phenomenon was unique to dematiaceous fungal species, including *A. niger* and *Stachybotrys chartarum*. Dematiaceous fungi are of particular interest from an occupational safety and public health standpoint as they are common contaminants of water damaged cellulose-based building materials. Examination of the literature indicated that previous attempts by other laboratories to produce MALDI-TOF MS spectra from *A. niger* had also been unsuccessful. Valentine and coworkers attempted to fingerprint *A. niger* in 2002 and stated in their manuscript, "*The analysis of some of the fungal samples, e.g., A. niger, were particularly challenging. A. niger was difficult to analyze by MALDI under all circumstances including pretreatment. Biomarkers were not easily detected for this fungal species... Further efforts to analyze this fungal species are needed.*" (*32*).

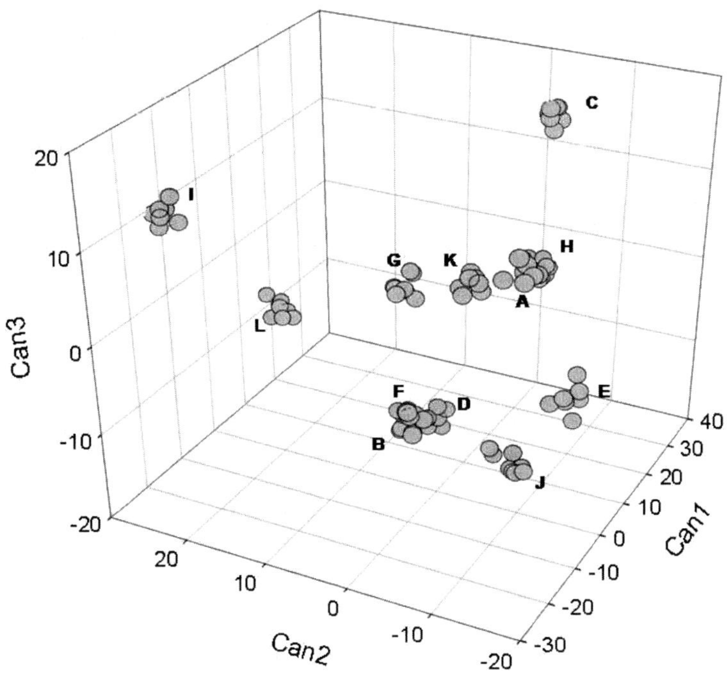

Figure 3. Three-dimensional canonical discriminant plot for MALDI-TOF MS data derived from Penicillium species from the whole fungal dataset. (A) P. aurantiogriseum; (B) P. brevicompactum; (C) P. citrinum; (D) P. chrysogenum; (E) P. expansum; (F) P. fellutanum; (G) P. jensenii; (H) P. melinii; (I) P. purpurogenum; (J) P. roqueforti; (K) P. simplicissimum; (L) P. variable.

Dematiaceous fungi contain eumelanins, a class of high molecular weight polymers of dihydroxynapthalene or dihydroxyphenylalanine, as a component of their cell wall (*41*). Melanins help provide stability and strength to the spore wall. Melanins are dark pigments that have photo- and chemo-protective properties (*42*, *43*) and may contribute to the virulence of several pathogenic fungi (*44*). Melanin absorbs UV radiation and dissipates the energy through ultrafast internal conversion. Based on this information, we hypothesized that fungal melanin in the cell extracts of dematiaceous fungi was interfering with the MALDI desorption/ionization process (*45*).

The exact chemical makeup of the pigment present in *A. niger* is still a matter of active research. Some investigators have dubbed this pigment "aspergillin" and suggested it is a polymer of high-molecular weight melanins (*46*). For the purposes of this discussion we shall refer to the black pigment isolated from *A. niger* using the generic term "melanin." In the present investigation, black pigment was purified from *A. niger* using a series of protease and strong acid digestions to remove all proteinaceous material. These treatments left behind structural melanin "ghosts". Addition of purified melanin to pure protein and peptide standards resulted in strong suppression of the MALDI-TOF MS $[M+H]^+$ signals for those standards in all cases. This observation strengthened the hypothesis that fungal melanin from *A. niger* suppressed MALDI ionization (*45*). We further demonstrate that blocking *A. niger* melanin synthesis by adding tricyclazole, a polyketide melanin pathway inhibitor, to the culture media (*47*) results in dramatically improved MALDI-TOF MS fingerprint mass spectra (Figure 5). Figure 5A presents the MALDI-TOF MS fingerprint mass spectrum of *A. niger* under conventional (and melanin producing) conditions. The MALDI-TOF MS fingerprint mass spectrum of melanin-deficient *A. niger* (Figure 5B), in contrast, yields a high number of very abundant *m/z* signals that may be used for discrimination.

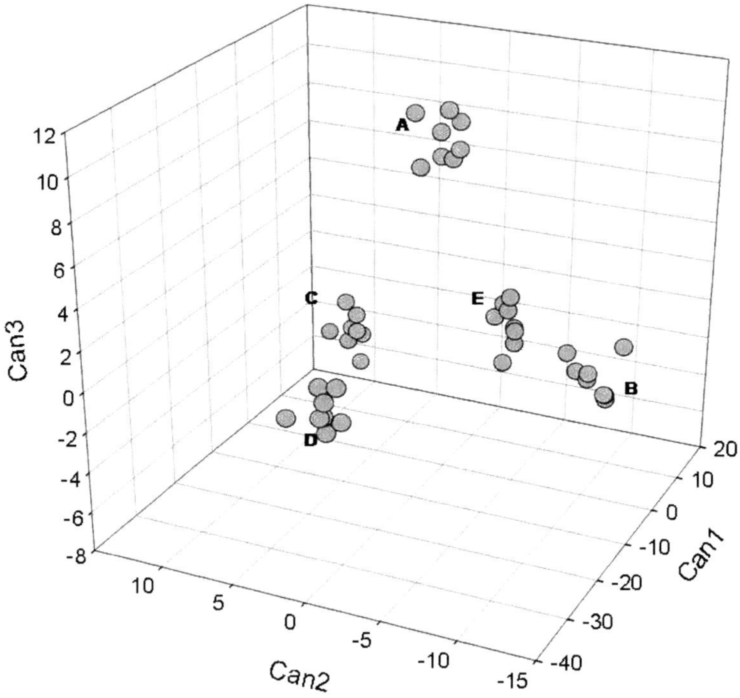

Figure 4. Three-dimensional canonical discriminant plot for MALDI-TOF MS data derived from five strains of Aspergillus flavus. (A) A. flavus NIOSH 15224; (B) A. flavus NIOSH 15417; (C) A. flavus ATCC 16883; (D) A. flavus NIOSH 34689; (E) A. flavus NIOSH PRC86N.

Conclusions

MALDI-TOF MS was used to generate a database of highly reproducible mass spectral "fingerprints" from 76 fungal isolates from 68 different species. Canonical discriminant analysis performed on the MALDI-TOF MS dataset was used to identify each species and/or strain with 98.5-100% accuracy, indicating that the methodology may be utilized for objective identification of fungi that complements traditional subjective identification techniques based on observation

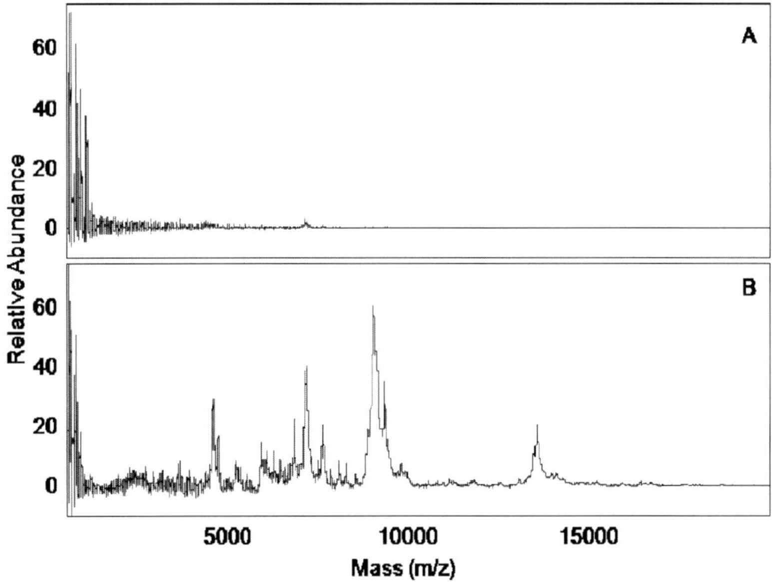

Figure 5. MALDI-TOF MS fingerprint mass spectrum of the dematiaceous fungus Aspergillus niger (A) grown on malt extract agar (B) grown on malt extract agar supplemented with 1% ethanol containing 50 µg/mL tricyclazole.

of colony morphology. In addition, we have confirmed that the presence of fungal melanins in the MALDI sample deposit results in significant suppression of the MALDI process, and consequently yields poor TOF mass spectra. The limitation of poor fingerprint mass spectra produced from dematiaceous fungi may be circumvented by judicious choice of culture conditions which prohibit melanin formation, such as using the polyketide melanin pathway inhibitor, tricyclazole. Blocking melanin formation allows high quality MALDI-TOF MS fingerprint data to be acquired from dematiaceous fungi such as *A. niger*.

References

1. Eduard, W. Fungal spores: a critical review of the toxicological and epidemiological evidence as a basis for occupational exposure limit setting. *Crit. Rev. Toxicol.* **2009**, *39*, 799–864.
2. Green, B. J.; Tovey, E. R.; Sercombe, J. K.; Blachere, F. M.; Beezhold, D. H.; Schmechel, D. Airborne fungal fragments and allergenicity. *Med. Mycol.* **2006**, *44* (Suppl 1), S245–55.
3. McGinnis, M. R. Indoor mould development and dispersal. *Med. Mycol.* **2007**, *45*, 1–9.
4. Fischer, G.; Dott, W. Relevance of airborne fungi and their secondary metabolites for environmental, occupational and indoor hygiene. *Arch. Microbiol.* **2003**, *179*, 75–82.
5. McGinnis, M. R. Pathogenesis of indoor fungal diseases. *Med. Mycol.* **2004**, *42*, 107–17.

6. Selman, M.; Lacasse, Y.; Pardo, A.; Cormier, Y. Hypersensitivity pneumonitis caused by fungi. *Proc. Am. Thorac. Soc.* **2010**, *7*, 229–36.
7. Black, P. N.; Udy, A. A.; Brodie, S. M. Sensitivity to fungal allergens is a risk factor for life-threatening asthma. *Allergy* **2000**, *55*, 501–4.
8. Downs, S. H.; Mitakakis, T. Z.; Marks, G. B.; Car, N. G.; Belousova, E. G.; Leuppi, J. D.; Xuan, W.; Downie, S. R.; Tobias, A.; Peat, J. K. Clinical importance of Alternaria exposure in children. *Am. J. Respir Crit. Care Med.* **2001**, *164*, 455–9.
9. Li, T.; Liu, B.; Chen, Y. Characterization of Aspergillus spores by matrix-assisted laser desorption/ionization time-of-flight mass spectrometry. *Rapid Commun. Mass Spectrom.* **2000**, *14*, 2393–2400.
10. O'Hollaren, M. T.; Yunginger, J. W.; Offord, K. P.; Somers, M. J.; O'Connell, E. J.; Ballard, D. J.; Sachs, M. I. Exposure to an aeroallergen as a possible precipitating factor in respiratory arrest in young patients with asthma. *N. Engl. J. Med.* **1991**, *324*, 359–63.
11. Charlier, C.; Lahoulou, R.; Dupont, B. Systemic mycosis in patients without evidence of immunosuppression. *J. Mycol. Med.* **2005**, *15*, 22–32.
12. Gugnani, H. C. Ecology and taxonomy of pathogenic Aspergilli. *Front. Biosci.* **2003**, *8*, s346–57.
13. Fischer, G.; Braun, S.; Dott, W. Profiles of microfungi--Penicillium chrysogenum and P. expansum. *Int. J. Hyg. Environ. Health.* **2003**, *206*, 65–7.
14. Frisvad, J. C.; Nielsen, K. F.; Samson, R. A. Recommendations concerning the chronic problem of misidentification of mycotoxigenic fungi associated with foods and feeds. *Adv. Exp. Med. Biol.* **2005**, *571*, 33–46.
15. Karas, M.; Hillenkamp, F. Laser desorption ionization of proteins with molecular masses exceeding 10000 daltons. *Anal. Chem.* **1988**, *60*, 2299–2301.
16. Karas, M.; Bachmann, D.; Bahr, U.; Hillenkamp, F. Matrix-assisted ultraviolet-laser desorption of nonvolatile compounds. *Int. J. Mass Spectrom.* **1987**, *78*, 53–68.
17. Cotter, R. J. The new time-of-flight mass spectrometry. *Anal. Chem.* **1999**, *71*, 445A–451A.
18. Claydon, M. A.; Davey, S. N.; Edwards-Jones, V.; Gordon, D. B. The rapid identification of intact microorganisms using mass spectrometry. *Nat. Biotechnol.* **1996**, *14*, 1584–1586.
19. Krishnamurthy, T.; Ross, P. L. Rapid identification of bacteria by direct matrix-assisted laser desorption/ionization mass spectrometric analysis of whole cells. *Rapid Commun. Mass Spectrom.* **1996**, *10*, 1992–1996.
20. Krishnamurthy, T.; Ross, P. L.; Rajamani, U. Detection of pathogenic and non-pathogenic bacteria by matrix-assisted laser desorption/ionization time-of-flight mass spectrometry. *Rapid Commun. Mass Spectrom.* **1996**, *10*, 883–888.
21. Welham, K. J.; Domin, M. A.; Scannell, D. E.; Cohen, E.; Ashton, D. S. The characterization of micro-organisms by matrix-assisted laser desorption/ionization time-of-flight mass spectrometry. *Rapid Commun. Mass Spectrom.* **1998**, *12*, 176–180.

22. Demirev, P. A.; Ho, Y. P.; Ryzhov, V.; Fenselau, C. Microorganism identification by mass spectrometry and protein database searches. *Anal. Chem.* **1999**, *71*, 2732–2738.
23. Fenselau, C.; Demirev, P. A. Characterization of intact microorganisms by MALDI mass spectrometry. *Mass Spectrom. Rev.* **2001**, *20*, 157–171.
24. Bright, J. J.; Claydon, M. A.; Soufian, M.; Gordon, D. B. Rapid typing of bacteria using matrix-assisted laser desorption ionization time-of-flight mass spectrometry and pattern recognition software. *J. Microbiol. Methods* **2002**, *48*, 127–138.
25. Pineda, F. J.; Antoine, M. D.; Demirev, P. A.; Feldman, A. B.; Jackman, J.; Longenecker, M.; Lin, J. S. Microorganism identification by matrix-assisted laser/desorption ionization mass spectrometry and model-derived ribosomal protein biomarkers. *Anal. Chem.* **2003**, *75*, 3817–3822.
26. Williams, T. L.; Andrzejewski, D.; Lay, J. O., Jr.; Musser, S. M. Experimental factors affecting the quality and reproducibility of MALDI TOF mass spectra obtained from whole bacteria cells. *J. Am. Soc. Mass Spectrom.* **2003**, *14*, 342–351.
27. Dworzanski, J. P.; Snyder, A. P.; Chen, R.; Zhang, H.; Wishart, D.; Li, L. Identification of bacteria using tandem mass spectrometry combined with a proteome database and statistical scoring. *Anal. Chem.* **2004**, *76*, 2355–2366.
28. Hettick, J. M.; Kashon, M. L.; Simpson, J. P.; Siegel, P. D.; Mazurek, G. H.; Weissman, D. N. Proteomic Profiling of Intact Mycobacteria by Matrix-Assisted Laser Desorption/Ionization Time-of-Flight Mass Spectrometry. *Anal. Chem.* **2004**, *76*, 5769–5776.
29. Hettick, J. M.; Kashon, M. L.; Slaven, J. E.; Ma, Y.; Simpson, J. P.; Siegel, P. D.; Mazurek, G. N.; Weissman, D. N. Discrimination of intact mycobacteria at the strain level: A combined MALDI-TOF MS and biostatistical analysis. *Proteomics* **2006**, *6*, 6416–6425.
30. Amiri-Eliasi, B.; Fenselau, C. Characterization of protein biomarkers desorbed by MALDI from whole fungal cells. *Anal. Chem.* **2001**, *73*, 5228–5231.
31. Welham, K. J.; Domin, M. A.; Johnson, K.; Jones, L.; Ashton, D. S. Characterization of fungal spores by laser desorption/ionization time-of-flight mass spectrometry. *Rapid Commun. Mass Spectrom.* **2000**, *14*, 307–310.
32. Valentine, N. B.; Wahl, J. H.; Kingsley, M. T.; Wahl, K. L. Direct surface analysis of fungal species by matrix-assisted laser desorption/ionization mass spectrometry. *Rapid Commun. Mass Spectrom.* **2002**, *16*, 1352–1357.
33. Chen, H. Y.; Chen, Y. C. Characterization of intact Penicillium spores by matrix-assisted laser desorption/ionization mass spectrometry. *Rapid Commun. Mass Spectrom.* **2005**, *19*, 3564–8.
34. Hettick, J. M.; Green, B. J.; Buskirk, A. D.; Kashon, M. L.; Slaven, J. E.; Janotka, E.; Blachere, F. M.; Schmechel, D.; Beezhold, D. H. Discrimination of Aspergillus isolates at the species and strain level by matrix-assisted laser desorption/ionization time-of-flight mass spectrometry fingerprinting. *Anal. Biochem.* **2008**, *380*, 276–81.

35. Hettick, J. M.; Green, B. J.; Buskirk, A. D.; Kashon, M. L.; Slaven, J. E.; Janotka, E.; Blachere, F. M.; Schmechel, D.; Beezhold, D. H. Discrimination of Penicillium isolates by matrix-assisted laser desorption/ionization time-of-flight mass spectrometry fingerprinting. *Rapid Commun. Mass Spectrom.* **2008**, *22*, 2555–2560.
36. Santos, C.; Paterson, R. R. M.; Venancio, A.; Lima, N. Filamentous fungal characterizations by matrix-assisted laser desorption/ionization time-of-flight mass spectrometry. *J. Appl. Microbiol.* **2010**, *108*, 375–385.
37. Bernardo, K.; Pakulat, N.; Macht, M.; Krut, O.; Seifert, H.; Fleer, S.; Hünger, F.; Krönke, M. Identification and discrimination of Staphylococcus aureus strains using matrix-assisted laser desorption/ionization-time of flight mass spectrometry. *Proteomics* **2002**, *2*, 747–753.
38. Jarman, K. H.; Daly, D. S.; Petersen, C. E.; Saenz, A. J.; Valentine, N. B.; Wahl, J. H. Extracting and visualizing matrix-assisted laser desorption/ionization time-of-flight mass spectral fingerprints. *Rapid Commun. Mass Spectrom.* **1999**, *13*, 1586–1594.
39. Wang, Y.; Aisen, P.; Casadevall, A. Melanin, melanin "ghosts," and melanin composition in Cryptococcus neoformans. *Infect. Immun.* **1996**, *64*, 2420–4.
40. Braga-Neto, U.; Hashimoto, R.; Dougherty, E. R.; Nguyen, D. V.; Carroll, R. J. Is cross-validation better than resubstitution for ranking genes? *Bioinformatics* **2004**, *20*, 253–8.
41. Dixon, D. M.; Polak-Wyss, A. The medically important dematiaceous fungi and their identification. *Mycoses* **1991**, *34*, 1–18.
42. Meredith, P.; Sarna, T. The physical and chemical properties of eumelanin. *Pigm. Cell Res.* **2006**, *19*, 572–94.
43. Meredith, P.; Riesz, J. Radiative relaxation quantum yields for synthetic eumelanin. *Photochem. Photobiol.* **2004**, *79*, 211–6.
44. Jacobson, E. S. Pathogenic roles for fungal melanins. *Clin. Microbiol. Rev.* **2000**, *13*, 708–17.
45. Buskirk, A. D.; Green, B. J.; Hettick, J. M.; Chipinda, I.; Law, B. F.; Siegel, P. D.; Slaven, J. E.; Beezhold, D. Fungal pigments inhibit the acquisition of MALDI-TOF mass spectra. Manuscript in preparation.
46. Ray, A. C.; Eakin, R. E. Studies on the biosynthesis of aspergillin by Aspergillus niger. *Appl. Microbiol.* **1975**, *30*, 909–15.
47. Yurlova, N. A.; de Hoog, G. S.; Fedorova, L. G. The influence of ortho- and para-diphenoloxidase substrates on pigment formation in black yeast-like fungi. *Stud. Mycol.* **2008**, *61*, 39–49.

Chapter 4

Fungal Metabolites for Microorganism Classification by Mass Spectrometry

Vladimir Havlicek[1,*] and Karel Lemr[2]

[1]Institute of Microbiology, Academy of Sciences of the Czech Republic, Videnska 1083, CZ 142 20 Prague 4, Czech Republic
[2]Faculty of Science, Palacky University, Tr. 17. Listopadu 12, 771 46 Olomouc, Czech Republic
***vlhavlic@biomed.cas.cz**

The current molecular, serological and emphasized mass spectral approaches used in fungal diagnostics are evaluated. An overview of low-molecular weight metabolites that have been used in medical mycology during either *in vitro* or *in vivo* experiments is presented. Fungal spores are suggested as a viable source of fungal biomarkers, as conidia inhalation is an important entrance gateway to host infection. In addition to MALDI (Matrix Assisted Laser Desorption/Ionization) typing, an alternative diagnostic method based on the knowledge of specific nonribosomal and other fungal metabolite structures is suggested. This approach can be used in fungal mixture analysis and represents a similar benefit to what we have learned from proteomics: going from peptide mapping to peptide sequencing could give better identification rates.

Introduction

Classical isolation of fungal pathogens for mycological evaluation is often an exercise in frustration, as blood cultures are notoriously insensitive, especially in the early stages of a disease (*1*). The current non–mass spectrometry high-throughput diagnostic tools have been based on RNA/DNA data, serological testing and/or arrays of automated biochemical tests.

The experimental approaches based on fungal oligonucleotide arrays can be fast (24 hours), sensitive (10 pg of yeast genomic DNA per assay) and specific

(97%) (*2*). Importantly, multiple fungal species that cause fungemia can also be detected (*3*). However, the success rates of these approaches are directly related to the limited number of available oligonucleotide probes in the corresponding library (usually hundreds of reference strains). In addition, the fungal nucleic acid must be isolated from the nucleic acid of the samples (blood or bronchoalveolar lavage fluid).

For other genotyping methods, see the recent review by Abdin et al. (*4*). The SeptiFast (LightCyclers SeptiFast, Roche Diagnostics, Mannheim, Germany) represents an example of a commercial version of a real-time PCR (Polymerase Chain Reaction) diagnostic set (*5*).

Immunological methods currently used in medical mycology are time- intensive, laborious, and disadvantageous in several other ways (*6*). Immunocompromised fungemic patients do not reliably produce antibodies, and the presence of an antibody does not precisely distinguish between colonization and infection. Furthermore, antifungal antibodies are often created by a healthy individual, which also hampers the decision process. For invasive aspergillosis, a commercialized ELISA (Enzyme-Linked ImmunoSorbent Assay)-based kit (Platelia test, Bio-Rad, Marnes-la-Coquette, France) is now commonly used for detection of circulating galactomannan in blood with improved sensitivity (0.5–1 ng mL^{-1}) and specificity over the previous latex agglutination test using the same monoclonal antibody (*7*). Knowledge of the fungal proteome has also resulted in the application of antibodies (*8*, *9*) in antifungal therapies. It has been shown that simultaneous administration of the antibody-based inhibitor Hsp90 (heat shock protein 90) with an antifungal drug (amphotericin B) results in a fourfold decrease in mortality in patients with invasive candidiasis (*10*).

The commercialized biochemical approaches are represented by BD Phoenix (*11*) (Becton Dickinson Diagnostic Systems, France) and VITEK-2 Smart Carrier Station or API/ID 32 (bioMerieux, Marcy L'Etoile, France) automated systems utilizing various panels and biochemical cards, respectively. For example, the *Neisseria-Haemophilus* 64-well identification card in the VITEK-2 system contains 30 biochemical tests in the following categories: 11 glycosidase and peptidase tests, 10 acidification tests, 5 alkalinization tests, and 4 miscellaneous tests. In a recent study, 91% of the bacterial strains (of the 188 strains tested) were correctly identified to the species level without additional tests within a 6-hour card incubation frame inside the VITEK-2 (*12*). Both Becton Dickinson and bioMerieux provide antibiotic susceptibility testing systems. The current databases, however, comprise just several hundred of the most common species of microorganisms, but include genera *Aspergillus*, *Candida*, and *Fusarium* and dimorphic fungi strains.

A comparison of MALDI-TOF MS (Matrix Assisted Laser Desorption/ Ionization-Time-of-Flight Mass Spectrometry) and biochemical approaches has been reported in a study on coagulase-negative staphylococci. The final percentages of correct results, misidentifications and absence of identification, respectively, were 97.4%, 1.3%, and 1.3% with MALDI-TOF MS, 79%, 21%, and 0% with the Phoenix, and 78.6%, 10.3%, and 0.9% with the VITEK-2 system (*13*). These results rationalize the recent acquisition of the AnagnosTec company (Potsdam-Golm, Germany) by bioMerieux and also the start of a joint bioMerieux-Shimadzu venture in May 2010 (*14*). Microorganism identification by MALDI typing beats the biochemical approaches not only in the higher identification rates. Mass spectral approach is also less expensive in consumables (<0.2 Euro/ID), it is faster and high throughput (>100 ID/hour). The sample preparation protocol is simpler and good for blood cultures and microbial detection in urine.

Typing Microorganisms by Mass Spectrometry

Mass spectrometry has been recognized as an indispensable molecular tool for clinical microbiologists, and it is particularly useful for high-throughput pathogen identification (*15*). In the growing bacterial research field, fiscal year 2009 was commercially successful for selected mass spectrometric vendors. This indicated better "penetration" for companies into otherwise less-accessible hospital laboratories; the medical community has begun losing its fear of the sometimes costly black-box mass spectrometry instrumentation (*13*).

MALDI-TOF MS–based bacterial typing represents a routine, fast and reliable method with reproducible and automated sample preparation protocols covering cultivation and standardized colony pick-up (*16*). Mycologists can assign non-fermenting bacteria (*17*), and the protocols allow for microorganism identification in less than 30 minutes once the blood culture is detected as positive (*18*). With the growth of microorganism databases, one of the remaining obstacles is represented by bacterial strain mix analysis. In fungal analysis, the second obstacle is the longer cultivation period mandatory for sufficient material production and visualization of fungal ribosomal proteins by MALDI typing.

Recently, a major health threat was caused by opportunistic (usually of the BSL-2 risk category) and resistant *Candida albicans*, *Cryptococcus neoformans* and *Aspergillus fumigatus* isolates (*19*). In addition, the spectrum of emerging life-threatening pathogens has increased significantly over the past two decades and has involved non-*albicans Candidi* and non-*fumigatus Aspergilli*. We also often encounter opportunistic yeast-like fungi (*Trichosporon* spp., *Rhodotorula* spp., and *Geotrichum capitatum)*; the zygomycetes; hyaline molds (such as *Fusarium*, *Acremonium*, *Scedosporium*, *Paecilomyces*, and *Trichoderma* species); and a wide variety of dematiaceous fungi (e.g., *Bipolaris*, *Exophiala*, *Phialophora*, and *Wangiella* spp.) (*20*). For diagnosis of the most common pathogens, at least some molecular/serological tests are commercially available (see the introductory section). For rare and emerging molds, however, detection tools are usually missing.

The considerable potential of the microbiological market, particularly in the field of MALDI strain typing, has been recognized by AnagnosTec (Potsdam-Golm, Germany), Bruker Daltonics (Bremen, Germany) and Waters (Manchester, UK), who have introduced the Saramis (*21*), the Biotyper (*17*) and the MicrobeLynx (*22*) solutions, respectively. Whereas the Waters product, utilizing a database with several thousand entries, was discontinued a few years ago, Biotyper and Saramis seem to have an increasing commercial potential. The Saramis databases contain more than 35,000 spectra of more than 2000 species and 500 genera (including fungi) at present (*23*), and the databases were recently acquired by bioMerieux. Biotyper version 2.0 contained more than 70,000 entries with about 3,300 unique reference strains as of August 2010. Particularly in bacteriology, mass spectral typing approaches are expected to replace Gram staining and biochemical identification in the near future (*24*).

The same shift could soon be possible in the fungal field. In a recent study, a total of 18 type collection strains and 267 recent clinical isolates of *Candida, Cryptococcus*, *Saccharomyces*, *Trichosporon*, *Geotrichum*, *Pichia*, and *Blastoschizomyces* spp. were identified (*25*). Starting with cells from single colonies, accurate species identification by MALDI-TOF MS was achieved for 247 of the clinical isolates (92.5%). The remaining 20 isolates required complementation of the reference database with spectra for the appropriate reference strains, which were obtained from type culture collections or identified by 26S rRNA gene sequencing. The absence of a suitable reference strain from the MALDI-TOF MS database was clearly indicated by log(score) values too low for the respective clinical isolates; i.e., no false-positive identifications occurred (*25*).

The most recent (and very first) report on filamentous fungi was presented by Marklein in Bonn in 2010 (*26*). Although the presented spectra were not of excellent quality, improved sample preparation protocols (*27*) will certainly yield searchable typing data.

Low-Molecular Weight Fungal Molecules Used *in Vitro* and *in Vivo*

Evolutionarily, fungi have developed multiple combat and pathogenicity strategies for propagation and survival. These strategies and mechanisms can be studied by imaging mass spectrometry (*28*), e.g., for direct visualization of metabolites secreted by combating microrganisms (*29*). The idea of using biologically active small molecules for microorganism identification is not new, and the approach benefits from the large amount of information that can be obtained from the fungal metabolome (*30*). Going from the gene and proteome levels down to ribosomal and nonribosomal products poses a problem in terms of sample complexity. However, the subtle differences in gene products among genetically similar strains that make their mutual discrimination impossible can be resolved by profiling secondary metabolites (*31*, *32*) that are not required for the growth of the fungus but have diverse functions and activities (*33*).

Mycotoxins represent just a single subgroup of secondary metabolites, but they have been recognized as important environmental (*34*) and clinical markers (*35*). The epidithiodioxopiperazine metabolite gliotoxin is detectable in the sera of patients suffering from invasive aspergillosis (*36*). In *in vitro* experiments, the toxin was detected in *A. fumigatus* (with 98% frequency), *A. niger* (56%), *A. terreus* (37%), and *A. flavus* (13%) culture filtrates. The weakest feature of gliotoxin is its low specificity; in addition to *Aspergilli*, this biomarker has been found in *Candida albicans* and *Pseudallescheria boydii*. The mean gliotoxin concentration found for isolates of patients that were colonized or had possible invasive infection was 5.30±0.69 μg/ml, whereas the mean concentration of gliotoxin for isolates of patients with probable or proven invasive aspergillosis was 7.97±1.12 μg/ml.

Interesting research has begun in the field of aflatoxin exposure in HIV-positive patients (*37*). It has been shown that HIV transmission frequency is positively associated with mycotoxin-contaminated maize consumption in Africa. The relation between cancer and food suggests that fumonisin contamination rather than aflatoxin is the most likely factor causing maize consumption to promote HIV (*38*). There has also been speculation in the past few years that various productive *Aspergilli* spp. growing in HIV-positive patients contribute to high levels of aflatoxins in the hosts. This general idea requires further research and, if proven, could have many therapeutic and diagnostic implications (*39*). A recent paper reported that mycotoxins can be detected in body fluids and human tissue from patients exposed to mycotoxin producing molds in the environment (*40*). Trichothecene levels varied in urine, sputum, and tissue biopsies (lung, liver, brain) from undetectable (<0.2 ppb) to levels up to 18 ppb. Aflatoxin levels from the same types of tissues varied from 1.0 to 5.0 ppb. Ochratoxins isolated in the same type of tissues varied from 2.0 ppb to > 10.0 ppb.

Another important *A. fumigatus* clinical marker is 2-pentylfurane, detected both *in vitro* and in human breath samples by solid phase micro extraction and GC/MS (*41*). It is a non-specific marker as it has been detected also in *Fusarium spp.*, *A. terreus*, *A. flavus*, *S. pneumoniae* and to a lesser extent in *A. niger*.

For the detection of candidemia, monitoring of *D*/*L*-arabinitol ratio has been found useful in the context of hematological neutropenia (*42*). It has, however, been stressed that other serological tests have yielded much higher sensitivity in disease determination.

Interesting applications of fungal siderophores have been reported recently, both *in vitro* and in clinical cases (*43*). Siderophores are secreted under iron stress to scavenge iron from host proteins like transferrin or ferritin (*44*). *N*-alpha-methyl coprogen B was recently detected in sputum samples from patients with cystic fibrosis complicated with scedosporiosis (*45*). The causative agent, *Scedosporium apiospermum*, belongs to a broader fungal complex called *Pseudallescheria boydii sensu lato,* which is an emerging set of pathogens causing fatal invasive infections in both immunocompromised and immunocompetent hosts. The limit of siderophore detection in bodily fluid has not yet been determined.

On the contrary, a semiquantitative information is known for selected metabolites present on the spores of *Pseudallescheria boydii,* a close relative to *S. apiospermum*. A set of putative non-ribosomal peptides named pseudacyclins has recently been found. Mass spectrometry indicated that there was 5×10^{-20} mol of pseudacyclin A on one CBS 119458 spore, which corresponds roughly to 30 000 molecules per spore (*46*). The presence of these peptides on inhaled fungal spores creates the possibility for exploitation of pseudacyclins as early indicators of fungal infections caused by *Pseudallescheria* species.

In general, nonribosomal peptides contain unique structural features, such as heterocyclic elements; D-amino acids; and glycosylated, fatty-acylated, hydroxylated, nitrated, and *N*-methylated residues. In contrast to proteins produced by ribosomal synthesis, nonribosomal products contain not only the common 21 coded amino acids but also hundreds of different building blocks. The huge complexity and diversity offered by cyclic peptides/depsipeptides provides us with a unique and extremely selective diagnostic tool (*47*). Most importantly, there is little evidence for the presence of nonribosomal synthetases in mammals. This fact led us to the original idea of using cyclic peptides/depsipeptides as specific markers of fungal infections. In other words, if there is a nonribosomal cyclic peptide marker found in a human/animal sample (tissue, blood, urine, etc.), it should serve as an indicator of a fungal, or possibly bacterial, infection of the particular host (*30*).

Similar to the use of pseudacyclins in *Pseudallescheria*, enniatin depsipeptides, linear acremostatins, cyclic paecilodepsipeptides, and trichosporins/trichoderins could be used for the assignment of infections caused by *Fusarium* (*48*), *Acremonium* (*49*), *Paecilomyces* (*50*), and *Trichoderma* (*51*, *52*), respectively. For various *Aspergilli* (including medically important species), multiple nonribosomal peptides have been predicted by polyketide and/or nonribosomal peptide synthetase gene sequencing (*53*), but just a few cyclic nonribosomal peptide siderophores have been characterized (*54*). Intact cyclic peptides then can be detected in bodily fluids directly or by various immunological methods based on these antigens (*55*). At present, there is no evidence supporting the existence of nonribosomal peptides in *Candidi*, *Histoplasma* and *Cryptococci*.

Although the potential of low-molecular weight fungal components for clinical diagnostics is not doubtful (*56*), one must stress that some secondary metabolites reported in this book chapter are experimental biomarkers. Some markers have been detected exclusively *in vitro* (cyclic peptides) and clinical studies are needed to evaluate the sensitivity of test procedures based on these biomarkers. Some markers have also been found *in vivo* (gliotoxin, aflatoxins, 2-methylfurane), however, their specificity with respect to fungal genus was low. The first paper on the application of siderophores has been published just recently, but the marker specificity still remains to be evaluated.

Concluding Remarks

The colonization of healthy individuals by various fungi is enormous (*57*). For pathogenic genera, including new and emerging fungi, fast, sensitive, reliable and early-stage diagnostic methods are lacking. Building a large microorganism database for MALDI typing will allow this technique to compete successfully with other molecular/serological/biochemical tools currently used. This particular task is being carried out by vendors having growing databases containing thousands of reference strains at present. The major obstacle—the determination of mixed microorganisms—can be resolved by the application of strain-specific low-molecular weight biomarkers, which could be implemented within the existing databases. This review aims at stimulating the companies to focus more on the well described marker molecules with known structure and evaluated specificity, which could give higher identification rates of pathogenic microbial strains.

Fungal spores seem to be an ideal source of fungal biomarkers, as conidia inhalation represents the most frequent entrance to the host body. In terms of specificity, the nonribosomal peptides represent important fungal biomarkers, although the structure elucidation of new representatives poses a considerable analytical problem (*58*, *59*). This is why one of the most promising products of non-ribosomal peptide synthetase (NRPS8) from *A. fumigatus* still remains to be discovered. Although predicted several years ago, the putative non-ribosomal product containing six amino acid residues has not yet been isolated (*53*). On the contrary, several mycotoxins and siderophores have been confirmed as potential biomarkers for selected infections caused by fungi.

For an initial rough assignment of known low-molecular weight candidates found during *in vitro* or *in vivo* experiments, public or commercial databases containing various natural products might be helpful (*60–65*). The absolute structures of new compounds, however, must be determined by advanced spectrometric and spectroscopic tools upon biomarker isolation.

Acknowledgments

This work was supported by the Ministry of Education, Youth and Sports of the Czech Republic (LC07017, ME10140, and MSM6198959216) and the Institutional Research Concept (AV0Z50200510). The authors do not have any associations that might pose a conflict of interest.

References

1. Thaler, M.; Pastakia, B.; Shawker, T. H.; Oleary, T.; Pizzo, P. A. *Ann. Int. Med.* **1988**, *108*, 88–100.
2. Leaw, S. N.; Chang, H. C.; Barton, R.; Bouchara, J. P.; Chang, T. C. *J. Clin. Microbiol.* **2007**, *45*, 2220–2229.
3. Hsiue, H. C.; Huang, Y. T.; Kuo, Y. L.; Liao, C. H.; Chang, T. C.; Hsueh, P. R. *Clin. Microbiol. Infect.* **2010**, *16*, 493–500.
4. Abdin, M. Z.; Ahmad, M. M.; Javed, S. *Arch. Microbiol.* **2010**, *192*, 409–425.
5. Donnelly, J. P. *Clin. Infect. Dis.* **2006**, *42*, 487–489.
6. Quindos, G. *Clin. Microbiol. Infect.* **2006**, *12*, 40–52.
7. Giacchino, M.; Chiapello, N.; Bezzio, S.; Fagioli, F.; Saracco, P.; Alfarano, A.; Martini, V.; Cimino, G.; Martino, P.; Girmenia, C. *J. Clin. Microbiol.* **2006**, *44*, 3432–3434.
8. Hauser, N.; Rupp, S.; Weber, A.; Tovar, G.; Hiller, E.; Borchers, K. Chip for diagnosing the presence of Candida based on thiol-specific antioxidant-like protein (TSA protein). EP 2007036352, 2007.
9. Carvalho, K. C.; Vallejo, M. C.; Camargo, Z. P.; Puccia, R. *Clin. Vaccine Immunol.* **2008**, *15*, 622–629.
10. Pachl, J.; Svoboda, P.; Jacobs, F.; Vandewoude, K.; van der Hoven, B.; Spronk, P.; Masterson, G.; Malbrain, M.; Aoun, M.; Garbino, J.; Takala, J.; Drgona, L.; Burnie, J.; Matthews, R. *Clin. Infect. Dis.* **2006**, *42*, 1404–1413.

11. Junkins, A. D.; Arbefeville, S. S.; Howard, W. J.; Richter, S. S. *J. Clin. Microbiol.* **2010**, *48*, 1929–1931.
12. Valenza, G.; Ruoff, C.; Vogel, U.; Frosch, M.; Abele-Horn, M. *J. Clin. Microbiol.* **2007**, *45*, 3493–3497.
13. Dupont, C.; Sivadon-Tardy, V.; Bille, E.; Dauphin, B.; Beretti, J. L.; Alvarez, A. S.; Degand, N.; Ferroni, A.; Rottman, M.; Herrmann, J. L.; Nassif, X.; Ronco, E.; Carbonnelle, E. *Clin. Microbiol. Infect.* **2010**, *16*, 998–1004.
14. Shimadzu and bioMérieux Enter into Partnership for Mass Spectrometry Applications in Microbiology. http://www.biomerieux.com/servlet/srt/bio/portail/dynPage?lang=en&doc=PRT_NWS_REL_G_PRS_RLS_225 (accessed Aug 18, 2010).
15. Ho, Y. P.; Reddy, P. M. *Clin. Chem.* **2010**, *56*, 525–536.
16. Demirev, P. A.; Fenselau, C. *Ann. Rev. Anal. Chem.* **2008**, *1*, 71–93.
17. Mellmann, A.; Cloud, J.; Maier, T.; Keckevoet, U.; Ramminger, I.; Iwen, P.; Dunn, J.; Hall, G.; Wilson, D.; LaSala, P.; Kostrzewa, M.; Harmsen, D. *J. Clin. Microbiol.* **2008**, *46*, 1946–1954.
18. Ferroni, A.; Suarez, S.; Beretti, J. L.; Dauphin, B.; Bille, E.; Meyer, J.; Bougnoux, M. E.; Alanio, A.; Berche, P.; Nassif, X. *J. Clin. Microbiol.* **2010**, *48*, 1542–1548.
19. Pihet, M.; Carrere, J.; Cimon, B.; Chabasse, D.; Delhaes, L.; Symoens, F.; Bouchara, J. P. *Med. Mycol.* **2009**, *47*, 387–397.
20. Pfaller, M. A.; Diekema, D. J. *J. Clin. Microbiol.* **2004**, *42*, 4419–4431.
21. Erhard, M.; Hipler, U. C.; Burmester, A.; Brakhage, A. A.; Woestemeyer, J. *Exp. Dermatol.* **2008**, *17*, 356–361.
22. Rajakaruna, L.; Hallas, G.; Molenaar, L.; Dare, D.; Sutton, H.; Encheva, V.; Culak, R.; Innes, I.; Ball, G.; Sefton, A. M.; Eydmann, M.; Kearns, A. M.; Shah, H. N. *Infect., Genet. Evol.* **2009**, *9*, 507–513.
23. SARAMIS FingerprintSpectra. http://www.anagnostec.eu/products-services/reference-databases/saramis-fingerprintspectra.html (accessed Aug 18, 2010).
24. Seng, P.; Drancourt, M.; Gouriet, F.; La Scola, B.; Fournier, P. E.; Rolain, J. M.; Raoult, D. *Clin. Infect. Dis.* **2009**, *49*, 543–551.
25. Marklein, G.; Josten, M.; Klanke, U.; Muller, E.; Horre, R.; Maier, T.; Wenzel, T.; Kostrzewa, M.; Bierbaum, G.; Hoerauf, A.; Sahl, H. G. *J. Clin. Microbiol.* **2009**, *47*, 2912–2917.
26. Marklein, G.; Muller, E. Oral presentation at the 3rd International Workshop on Scedosporium Meeting, Bonn, 6−8 May, 2010.
27. Sulc, M.; Peslova, K.; Zabka, M.; Hajduch, M.; Havlicek, V. *Int. J. Mass Spectrom.* **2009**, *280*, 162–168.
28. Esquenazi, E.; Yang, Y. L.; Watrous, J.; Gerwick, W. H.; Dorrestein, P. C. *Nat. Prod. Rep.* **2009**, *26*, 1521–1534.
29. Watrous, J.; Hendricks, N.; Meehan, M.; Dorrestein, P. C. *Anal. Chem.* **2010**, *82*, 1598–1600.
30. Nedved, J.; Sulc, M.; Jegorov, A.; Giannakopulos, A.; Havlicek, V. In *Clinical Proteomics, from Diagnosis to Therapy*; Eyk, J. E. V., Dunn, M. J., Eds.; Wiley-VCH Verlag GmbH: 2008; pp 483−509.

31. Jegorov, A.; Paizs, B.; Kuzma, M.; Zabka, M.; Landa, Z.; Sulc, M.; Barrow, M. P.; Havlicek, V. *J. Mass Spectrom.* **2004**, *39*, 949–960.
32. Jegorov, A.; Paizs, B.; Zabka, M.; Kuzma, M.; Havlicek, V.; Giannakopulos, A. E.; Derrick, P. J. *Eur. J. Mass Spectrom.* **2003**, *9*, 105–116.
33. Ishigami, K.; Katsuta, R.; Shibata, C.; Hayakawa, Y.; Watanabe, H.; Kitahara, T. *Tetrahedron* **2009**, *65*, 3629–3638.
34. Lau, A. P. S.; Lee, A. K. Y.; Chan, C. K.; Fang, M. *Atm. Environ.* **2006**, *40*, 249–259.
35. Sugui, J. A.; Pardo, J.; Chang, Y. C.; Zarember, K. A.; Nardone, G.; Galvez, E. M.; Muellbacher, A.; Gallin, J. I.; Simon, M. M.; Kwon-Chung, K. J. *Eukaryot. Cell* **2007**, *6*, 1562–1569.
36. Kupfahl, C.; Michalka, A.; Lass-Floerl, C.; Fischer, G.; Haase, G.; Ruppert, T.; Geginat, G.; Hof, H. *Int. J. Med. Microbiol.* **2008**, *298*, 319–327.
37. Jolly, P.; Jiang, Y.; Preko, P.; Wang, J. S.; Ellis, W.; Phillips, T.; Williams, J.; Besssler, P.; Baidoo, J.; Stiles, J. *Epidemiology* **2009**, *20*, S239–S239.
38. Williams, J. H.; Grubb, J. A.; Davis, J. W.; Wang, J. S.; Jolly, P. E.; Ankrah, N. A.; Ellis, W. O.; Afriyie-Gyawu, E.; Johnson, N. M.; Robinson, A. G.; Phillips, T. D. *Am. J. Clin. Nutr.* **2010**, *92*, 154–160.
39. Leema, G.; Kaliamurthy, J.; Geraldine, P.; Thomas, P. A. *Mol. Vision* **2010**, *16*, 843–854.
40. Hooper, D. G.; Bolton, V. E.; Guilford, F. T.; Straus, D. C. *Int. J. Mol. Sci.* **2009**, *10*, 1465–1475.
41. Syhre, M.; Scotter, J. M.; Chambers, S. T. *Med. Mycol.* **2008**, *46*, 209–215.
42. Arendrup, M. C.; Bergmann, O. J.; Larsson, L.; Nielsen, H. V.; Jarlov, J. O.; Christensson, B. *Clin. Microbiol. Infect.* **2010**, *16*, 855–862.
43. Bertrand, S.; Larcher, G.; Landreau, A.; Richomme, P.; Duval, O.; Bouchara, J. P. *BioMetals* **2009**, *22*, 1019–1029.
44. Kornitzer, D. *Curr. Opin. Microbiol.* **2009**, *12*, 377–383.
45. Bertrand, S.; Bouchara, J. P.; Venier, M. C.; Richomme, P.; Duval, O.; Larcher, G. *Med. Mycol.* **2010**, in press.
46. Pavlaskova, K.; Nedved, J.; Kuzma, M.; Zabka, M.; Sulc, M.; Sklenar, J.; Novak, P.; Benada, O.; Kofronova, O.; Hajduch, M.; Derrick, P. J.; Lemr, K.; Jegorov, A.; Havlicek, V. *J. Nat. Prod.* **2010**, *73*, 1027–1032.
47. Jegorov, A.; Haiduch, M.; Sulc, M.; Havlicek, V. *J. Mass Spectrom.* **2006**, *41*, 563–576.
48. Uhlig, S.; Ivanova, L.; Petersen, D.; Kristensen, R. *Toxicon* **2009**, *53*, 734–742.
49. Degenkolb, T.; Heinze, S.; Schlegel, B.; Strobel, G.; Grafe, U. *Biosci. Biotechnol. Biochem.* **2002**, *66*, 883–886.
50. Isaka, M.; Palasarn, S.; Lapanun, S.; Sriklung, K. *J. Nat. Prod.* **2007**, *70*, 675–678.
51. Iwatsuki, M.; Kinoshita, Y.; Niitsuma, M.; Hashida, J.; Mori, M.; Ishiyama, A.; Namatame, M.; Nishihara-Tsukashima, A.; Nonaka, K.; Masuma, R.; Otoguro, K.; Yamada, H.; Shiomi, K.; Omura, S. *J. Antibiot.* **2010**, *63*, 331–333.

52. Pruksakorn, P.; Arai, M.; Kotoku, N.; Vilcheze, C.; Baughn, A. D.; Moodley, P.; Jacobs, W. R.; Kobayashi, M. *Bioorg. Med. Chem. Lett.* **2010**, *20*, 3658–3663.
53. Cramer, R. A.; Stajich, J. E.; Yamanaka, Y.; Dietrich, F. S.; Steinbach, W. J.; Perfect, J. R. *Gene* **2006**, *383*, 24–32.
54. Schrettl, M.; Bignell, E.; Kragl, C.; Sabiha, Y.; Loss, O.; Eisendle, M.; Wallner, A.; Arst, H. N.; Haynes, K.; Haas, H. *PLoS Pathog.* **2007**, *3*, 1195–1207.
55. Safarcik, K.; Brozmanova, H.; Bartos, V.; Jegorov, A.; Grundmann, M. *Clin. Chim. Acta* **2001**, *310*, 165–171.
56. Urusov, A. E.; Zherdev, A. V.; Dzantiev, B. B. *Appl. Biochem. Microbiol.* **2010**, *46*, 253–266.
57. Ghannoum, M. A.; Jurevic, R. J.; Mukherjee, P. K.; Cui, F.; Sikaroodi, M.; Naqvi, A.; Gillevet, P. M. *PLoS Pathog.* **2010**, *6*, 8.
58. Liu, W. T.; Ng, J.; Meluzzi, D.; Bandeira, N.; Gutierrez, M.; Simmons, T. L.; Schultz, A. W.; Linington, R. G.; Moore, B. S.; Gerwick, W. H.; Pevzner, P. A.; Dorrestein, P. C. *Anal. Chem.* **2009**, *81*, 4200–4209.
59. Ng, J.; Bandeira, N.; Liu, W. T.; Ghassemian, M.; Simmons, T. L.; Gerwick, W. H.; Linington, R.; Dorrestein, P. C.; Pevzner, P. A. *Nat. Methods* **2009**, *6*, 596–U65.
60. Novel Antibiotics DataBase. http://www.nih.go.jp/~jun/NADB/search.html (accessed Aug 18, 2010).
61. Dunkel, M.; Fullbeck, M.; Neumann, S.; Preissner, R. *Nucl. Acids Res.* **2006**, *34*, D678–D683.
62. Smith, C. A.; O'Maille, G.; Want, E. J.; Qin, C.; Trauger, S. A.; Brandon, T. R.; Custodio, D. E.; Abagyan, R.; Siuzdak, G. *Ther. Drug Monit.* **2005**, *27*, 747–751.
63. Fahy, E.; Sud, M.; Cotter, D.; Subramaniam, S. *Nucleic Acids Res.* **2007**, *35*, W606–W612.
64. Rosenstein, I. J. *J. Chem. Educ.* **2005**, *82*, 652–654.
65. Nielsen, K. F.; Smedsgaard, J. *J. Chromatogr., A* **2003**, *1002*, 111–136.

Chapter 5

Rapid Profiling of Recombinant Protein Expression from Crude Cell Cultures by Matrix-Assisted Laser Desorption/Ionization Mass Spectrometry (MALDI-MS)

Scott C. Russell[*]

Department of Chemistry, California State University, Stanislaus One University Circle, Turlock, CA 95382
[*]srussell@chem.csustan.edu

This chapter focuses on the detection of recombinant proteins by matrix-assisted laser desorption/ionization performed directly on cell culture samples with minimal cleanup. The methods highlighted in this chapter offer the distinct advantage of speed compared to conventional approaches to characterizing recombinant proteins. This increased analysis speed opens the possibility for high throughput screening of recombinant protein expression, which should aid those looking to optimize recombinant protein expression conditions. Several approaches are highlighted including protein mass matching, monitoring expression dynamics, and bottom up proteomics for confident recombinant protein identification. Methods for detection of recombinant proteins directly from both prokaryotic and eukaryotic expression systems are summarized.

Introduction

Recombinant Protein Expression

The goal of this chapter is to highlight the advantages of using matrix-assisted laser desorption/ionization mass spectrometry (MALDI-MS) to quickly monitor recombinant protein expression directly from cell cultures. Recombinant protein expression in prokaryotic microorganisms generally involves either plasmid or viral cloning vectors (*1*). Plasmids are not integrated into the microorganisms

DNA and require selective pressure to maintain plasmid presence in the host cells (*1*). By contrast, the recombinant DNA of a viral a vector is integrated into the DNA of the host cell (*1*). Regardless of the vector chosen, the goal of recombinant protein expression is to utilize the host's genetic machinery to produce therapeutically relevant proteins (*1*). Recombinant protein expression was first realized in 1977 with the production of mammalian somatostatin in *E. coli* cells (*2*). Recombinant protein expression was then used commercially by Genentech to produce human insulin in *E. coli* cells (*3*). Since then, a large number of recombinant proteins have been produced via expression in *E. coli* (*4–6*).

Often times prokaryotic expression systems are not sufficient for the production of mammalian proteins (*7*). Problems can occur due to the lack of post translational modifications (PTM's) when working in prokaryotic expression systems (*7*). Consequently, it is often necessary to utilize eukaryotic expression systems to ensure proper protein folding and function (*7*). Additionally, expression conditions require optimization including vector choice, induction conditions and expression time (*4–6*).

Conventional Recombinant Protein Characterization

Unfortunately, conventional methods available to monitor the expression of recombinant proteins in microorganisms involve lengthy procedures that can take hours to days to complete (*8–10*). Since the timescale of analysis is on the order of the timescale for protein expression itself, real-time monitoring of protein expression by conventional methods is not possible. The inability to monitor protein expression in real time hinders the optimization of experimental parameters for abundant protein expression.

The most frequently used technique to monitor recombinant protein expression has been sodium dodecyl sulfate polyacrylamide gel electrophoresis (SDS-PAGE) (*8–10*). This method provides a visual representation of expressed proteins separated by molecular weight. Often times protein extracts cannot be fully resolved by molecular weight alone. Therefore, isoelectric separation is often carried out prior to running the gel, yielding two dimensional SDS-PAGE. The overexpression of a recombinant protein is readily observable via 2D-SDS-PAGE by comparing the expression profile of a control cell line with that of a cell line that has been genetically modified to express a recombinant protein. However, the molecular weight measured by gel electrophoresis is only approximate and full protein identification requires mass spectrometry (*11–18*). Additionally, post-translational modifications (PTM's) can be missed due to the poor mass accuracy and low resolution associated with gel electrophoresis (*8*). Further confidence in the recombinant protein identity can be carried out via antibody methods such as a Western blot or Enzyme-linked immunosorbent assay (ELISA) (*8*). However, these methods require significant additional analysis time as well as the potential for non-specific antibody interactions (*8*).

Bacteria Mass Spectrometry (A Brief History)

Mass spectrometry was first used to directly analyze bacteria in 1975 by gentle heating followed by electron impact ionization (*19*). Many of the methods to follow involved pyrolysis mass spectrometry performed directly on cells to generate lipid biomarkers (*20*, *21*). Fast atom bombardment was also successfully utilized to type species by detecting phospholipids directly from microorganisms (*22*). Proteins were later detected from microorganisms via electrospray ionization (ESI) (*23*). However, ESI is prone to clogging and produces spectra with multiple charge states, making filtration and fractionation necessary prior to analysis of proteins from microorganisms (*24–26*).

Alternatively, matrix-assisted laser desorption/ionization (MALDI) is more tolerant to salt contamination, produces simpler singly charged spectra, and does not require filtration or fractionation. These attributes have made MALDI the mass spectrometric method of choice for protein analysis directly from microorganisms (*27–30*). Often, cells or growth media are spotted directly on the MALDI plate along with organic acids such as acetic acid (*31*), trifluoroacetic acid (*32–34*), formic acid (*35*, *36*), or solvents such as ethanol (*36*), or methanol (*35*, *37*, *38*) and allowed to dry. These solvents and organic acids have been shown to selectively release and solubilize a subset of proteins, which improves spectral quality (*39*). A great deal of work has confirmed the identity of the major peaks generated from whole cell MALDI-MS as being translated proteins (*40–43*). MALDI matrices that have been preferred for protein detection directly from microorganisms are sinapinic acid (*37*, *44*), and ferulic acid (*45*) for high masses, and α-cyano-4-hydroxycinnamic acid for low masses (*28*).

The goal of the detection of proteins directly from microorganisms has largely been to use them as biomarkers for rapid species identification. While fingerprinting methods have shown promise for species identification (*46*, *47*), the advantage of using a proteomics approach is that the proteins in the whole cell MALDI mass spectrum are identified (*48–52*). Determining the protein's identity makes the proteomics approach intrinsically adaptable to the characterization of recombinant proteins via whole cell MALDI (*48–52*). With that in mind, proteins can be identified and characterized via whole-cell MALDI by intact protein mass matching (*48–50*), top down sequencing (*53*, *54*), bottom-up peptide mass matching following a residue specific digestion (*51*), or from peptide microsequences following a residue specific digestion and tandem MS (*52*).

Characterization of recombinant proteins via whole cell MALDI has direct applications in the biotech industry such as monitoring cell cultures for recombinant protein expression, viral or plasmid vector incorporation, PTM characterization, etc. The remainder of this chapter will highlight achievements in detection and identification of recombinant proteins directly from crude cell cultures via MALDI-MS. Additionally, insights into the appropriate experimental conditions necessary to achieve success in monitoring recombinant protein expression in crude cell cultures via MALDI-MS will be summarized.

Detection of Recombinant Proteins from Prokaryotic Microorganisms (Plasmid Vectors)

Intact Protein Detection (Mass Matching)

A personal communication by *Chait* is cited by two papers reporting the first detection of recombinant proteins via whole-cell MALDI-MS (*55*, *56*). In 1996 the first result demonstrating the ability of whole-cell MALDI to detect a plasmid-borne recombinant protein was published by Parker *et. al.* (*56*). This paper demonstrated that by performing MALDI-TOF directly on *E. coli* (strain DH5) cells, an HIV-1IIIB recombinant protein (56kDa) could be detected. The authors reported a whole cell MALDI-TOF mass spectrum of cells induced via IPTG with a low intensity peak observed matching the recombinant protein mass (*56*). The low signal intensity may reflect the use of α-cyano-4-hydroxycinnamic acid as the MALDI matrix, which is well suited to lower mass proteins, whereas sinapinic acid is preferred for higher mass proteins (*57–63*). Additionally, a low intensity peak was observed that matched a doubly charged ion of the recombinant protein. The authors noted the difficulty in observing the recombinant protein in the cell lysate. Cell lysis has been shown as something to avoid and has been shown to suppress protein signal, while selective solubilization of the proteins of interest is a preferred approach to improve MALDI mass spectral quality from whole cells (*30*).

Whole cell MALDI-TOF was later applied to the detection of multiple recombinant proteins with much improved spectral quality (*57*). Four recombinant proteins were studied; all of which were plasmid-borne. Three of these proteins, hTFIIB-NTD, *Pf*TFB-NTD, and *PF*Rd, would be considered low mass (~6 kDa) and one, a MalE/MerP fusion protein, would be considered higher mass (~ 50 kDa). The approach to the low mass protein analysis involved centrifugation of the cells and removal of the culture broth. The cell pellet was then resuspended in a minimal volume (10-15 μL) of a 50:50 mixture of water/acetonitrile. One to two μL of this suspension was then spotted with an equal volume of saturated sinapinic acid matrix solution and allowed to air dry (*57*). The high mass protein analysis involved sonication of the culture broth prior to mixing 10 μL of the culture broth with an equal volume of saturated sinapinic acid in 50:50 water/acetonitrile. This mixture was then spotted on the MALDI plate for mass analysis. It was speculated that the sonication aided in solubilizing the high mass fusion protein (*57*). Figure 1 shows the MALDI-TOF mass spectrum of the MalE/MerP fusion protein obtained directly from the sonicated cell culture (*57*).

Direct monitoring of recombinant protein expression from individual bacteria colonies selected from agar plates was later reported (*58*). Ten colonies were randomly lifted from an agar plate with 1-uL loops and placed in 50-uL of 50 % ACN, 0.1 % TFA. These cell suspensions were mixed vigorously and spotted for MALDI-TOF analysis using the sandwich method (*64*) with sinapinic acid (0.1M in 1:1:1 ACN, MeOH, water). The resulting MALDI mass spectra showed peaks at *m/z* 15830 – 15848 (predicted *m/z* = 15,849) with S/N ranging from 36 – 471. Figure 2a shows a mass spectrum from a single randomly picked colony that was expressing this protein (penvA), while Figure 2b shows a mass spectrum

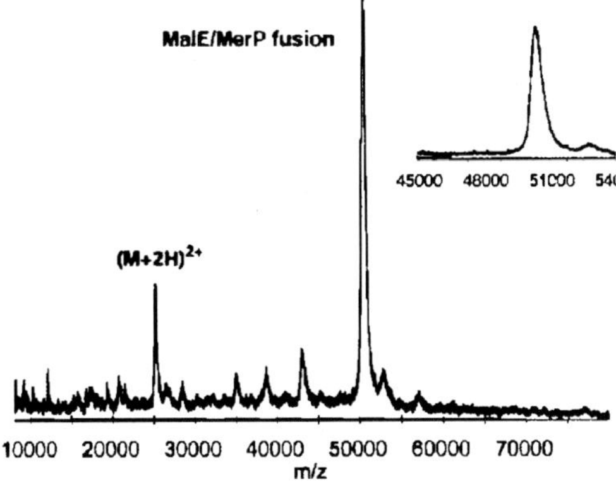

Figure 1. MALDI-TOF mass spectrum of a MalE/MerP fusion protein obtained from a sonicated cell culture. Reproduced with permission from reference (57). Copyright 1998, American Chemical Society.

of a colony not expressing the penvA protein (*58*). One can see a peak matching that of the penvA recombinant protein in the cells overexpressing it, but not the control cell line. These results demonstrate the ability of MALDI-TOF to monitor individual cultures for successful recombinant protein expression. This method would be of great utility to validate the selection of a colony from an agar plate prior to scaling up the culture.

Sample Cleanup

The use of a C_{18} resin cleanup/concentration step has been shown to further improve the MALDI-TOF mass spectral quality of recombinant proteins from crude cell samples (*59*). This approach resulted in the detection of one intracellular recombinant protein (glucosidase, ~ 50 kDa), and three extracellular recombinant proteins (4 kDa – 25 kDa). While this method does require some cleanup time, it is minimal and produced significant signal enhancement. Figures 3 shows MALDI-TOF mass spectra in which signal was greatly enhance for LCI (~7 kDa) and glucosidase (~53 kDa) (*59*).

Monitoring Recombinant Protein Expression Dynamics

It is important to note that unintended changes in endogenous protein expression may occur upon an induction event (*60*). This was characterized by MALDI-TOF in which whole cell spectra were monitored for changes following induction with IPTG (*60*). This study also reported the ability to monitor whole *E. coli* cells response to IPTG induction and detected overexpressed proteins at ~ 48 kDa (Rac protein), ~43 kDa (methyl transferase), and ~49 kDa (methyl transferase) (*60*). Figure 4 shows mass spectra taken at one, two, and three hours after induction of a 48 kDa Rac protein. This time-course study showed increased protein abundance at three hours after induction as opposed to one or two hours (*60*). This again, illustrates the power of using whole-cell MALDI to optimize growth conditions for optimal recombinant protein expression. All spectra were obtained by taking a 2 μL aliquot of an *E. coli* cell suspension and mixing with 18 μL of a saturated sinapinic acid solution in 50 % ACN, 2.5 % formic acid and spotted for MALDI-TOF analysis.

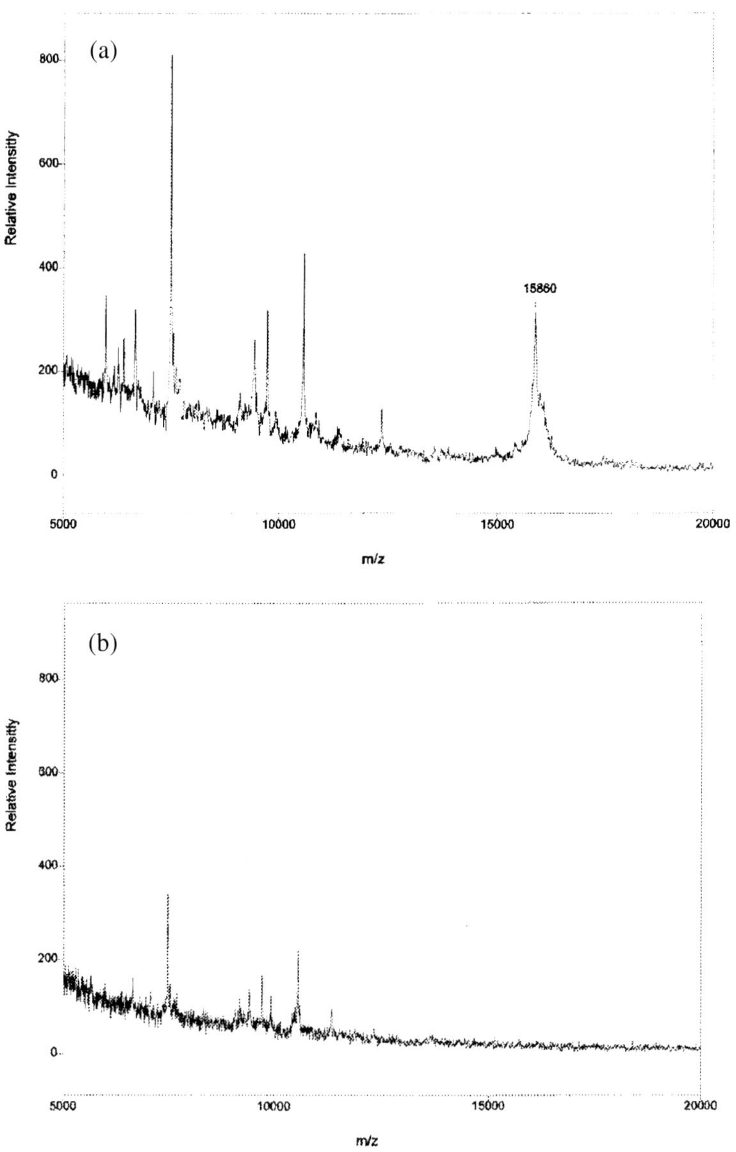

Figure 2. MALDI-TOF mass spectra from (a) a randomly selected E. coli bacterial clone expressing recombinant protein (penV) and (b) an E. coli bacterial clone not expressing recombinant protein (penV). Reproduced with permission from reference (58). Copyright 2000, BioTechniques.

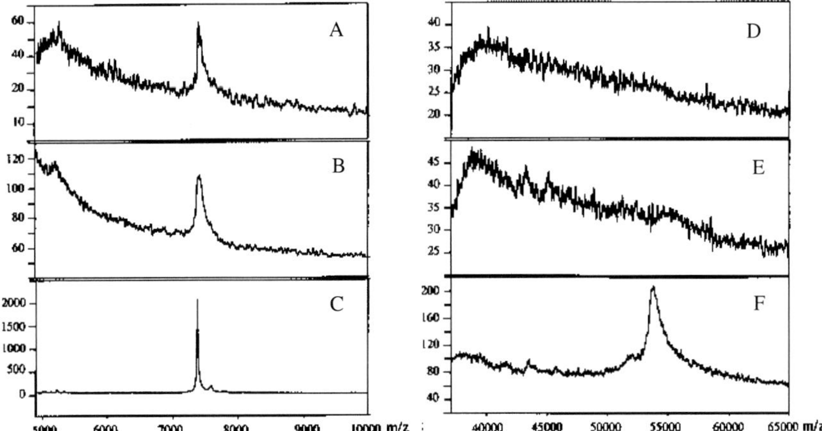

Figure 3. MALDI-TOF mass spectra of E. coli cultures expressing recombinant protein (LCI) with (A) direct sample analysis, (B) TFA-rinsed (C) C_{18} microcolumn purified. Also shown are MALDI-TOF mass spectra of E. coli cultures expressing recombinant protein (Glucosidase) with (D) direct sample analysis, (E) TFA-rinsed (F) C_{18} microcolumn purified. Reproduced with permission from reference (59). Copyright 2001, Elsevier Science Inc.

In another study, the overexpression of native green fluorescent protein (GFP, ~27 kDa) and GFP(histidine)$_6$ (~28kDa) was monitored over time following arabinose induction via whole cell MALDI-TOF (*65*). The fluorescence and MALDI-TOF signal for GFP were found to be directly correlated with one another. The fluorescence measurement showed sensitivity of two orders of magnitude greater than that of MALDI. However, MALDI has the distinct advantage of not requiring a protein specific fluorophore for detection. The MALDI sample preparation procedure was more elaborate than those described above. The procedure involved a 30 minute centrifugation step to concentrate the *E. coli* cells, which were then resuspended in MALDI matrix, vortexed and spotted on the MALDI plate (*65*). The authors reported using 2.5 M 3,5-dihydroxybenzoic acid in 90%MeOH/7.5% TFA (*65*), which was likely a typo evidenced by their citation of 2,5 DHB as a preferred matrix for reproducible whole cell MALDI spectra (*66*). However, the spectral quality shown was somewhat noisy in the range of the detected recombinant proteins (25kDa – 32kDa). This may have been improved by moving to sinapinic acid, which has shown greater success for MALDI of higher mass recombinant proteins directly from microorganisms (*57–61*).

Monitoring the Effect of Induction Method

In a 2006 study, a strong argument was made for the use of whole cell MALDI-TOF as a combinatorial method to quickly vary/optimize recombinant protein expression parameters (*61*). *E. coli* cells were monitored for recombinant glutathione S transferase expression via whole cell MALDI-TOF (*61*). Approximately 10^6 cells were spotted on the MALDI plate and mixed with

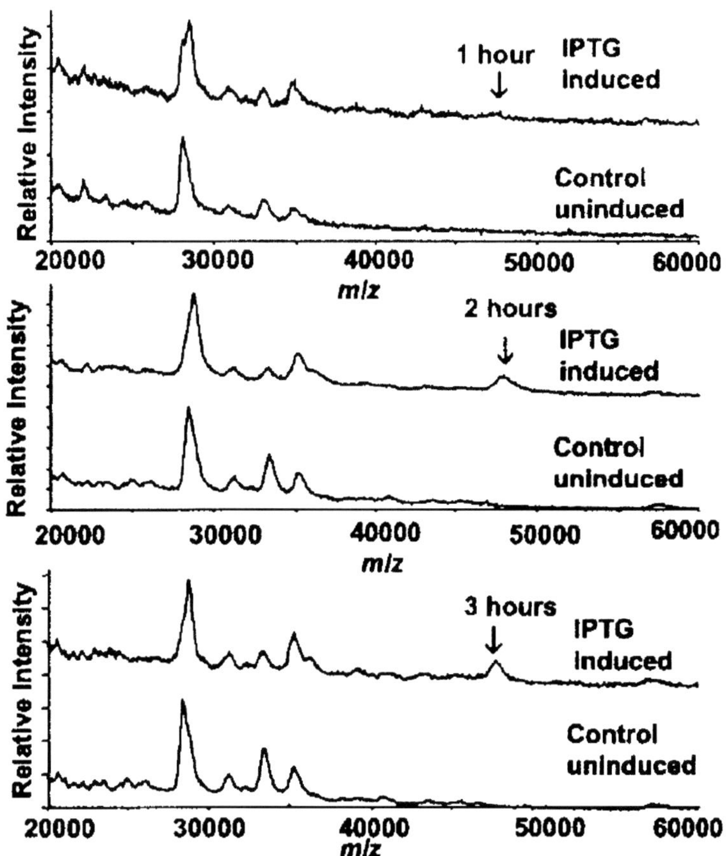

Figure 4. MALDI-TOF time course monitoring of IPTG-induced Rac protein expression from whole cells. Reproduced with permission from reference (60). Copyright 2002, Elsevier Science-USA.

sinapinic acid (10mg/mL in 30% ACN and 0.2% TFA) by the dried droplet method. Cells were monitored following arabinose induction and an intense peak at *m/z* 28,889 was observed, which matched that of the recombinant protein. The effect of using glucose as an alternative inducer was also studied to optimize protein expression. Induction parameters were also varied and the expression results were monitored (*61*). The speed of whole cell MALDI enables this combinatorial approach, while conventional methods do not (*61*).

Detection of Recombinant Proteins from Prokaryotic Microorganisms (Viral Vectors)

Intact Protein Detection (Mass Matching)

An alternative method to the use of plasmids is to use viral vectors for recombinant protein expression in bacteria (*1*). The following papers have

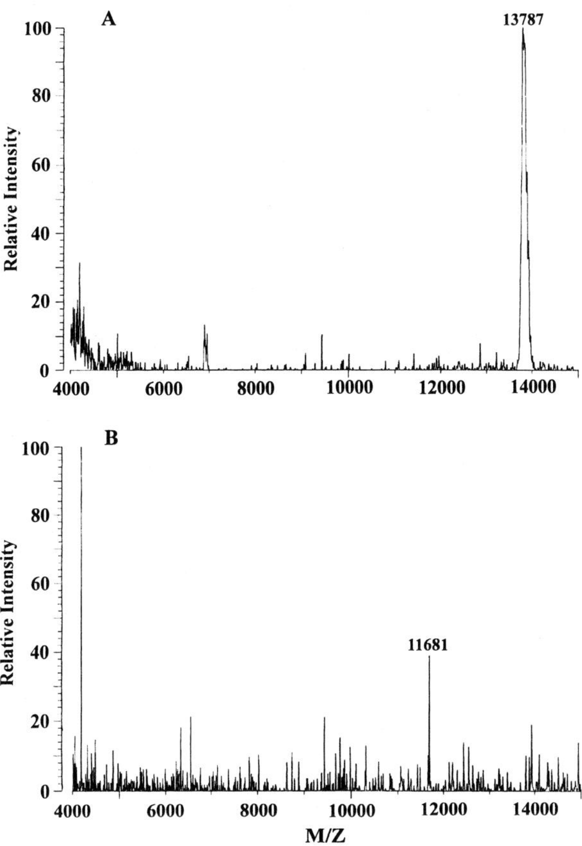

Figure 5. MALDI-TOF mass spectra of (A) MS2 bacteriophage in an aliquot of culture broth and (B) an aliquot of virus free culture broth. Reproduced with permission from reference (31). Copyright 1998, American Chemical Society.

demonstrated the ability of whole cell MALDI to detect viral-borne proteins in crude cell culture media and intact host cells. The first attempt at viral protein detection via whole-cell MALDI-TOF was performed in 1998 (*31*). This study involved the infection of host *E. coli* cells with bacteriophage MS2, which contains 180 copies of a coat protein (13,728 Da) that surround viral RNA (*67*). The authors demonstrated that this protein could be detected directly from the culture broth of infected *E. coli* cells. Figure 5a shows a mass spectrum of an aliquot of crude culture broth from *E. coli* cells infected with bacteriophage MS2 as well as uninfected *E. coli* cells crude culture broth as a control (Figure 5b). A peak at *m/z* 13,787 can clearly be seen in the spectrum of the viral infected sample, which is lacking in the spectrum from the uninfected sample. Acetic acid at a level above 10 % in the sinapinic acid matrix solution was found to be critical to detection of the viral protein. This matrix solution was mixed with the crude culture aliquots and spotted on the MALDI plate and allowed to dry. This method required only 3 minutes from sampling to mass analysis.

Detection of Recombinant Proteins from Eukaryotic Microorganisms (Plasmid & Viral Vectors)

Intact Protein Detection (Mass Matching)

It is often necessary to express recombinant proteins in eukaryotic microorganisms such as yeast (*1*). Many mammalian proteins require post-translational modifications (PTM's) that do not occur in prokaryotic cell lines (*7*). A lack of PTM's can lead to misfolded proteins that are often insoluble and lack their proper function. Characterization of plasmid-borne recombinant proteins expressed in eukaryotic cells has been demonstrated by MALDI-TOF without sample cleanup (*60*, *68*).

The first example involved detection of a plasmid-borne secreted recombinant protein expressed in *Pichia pastoris* (*68*). *P. pastoris* cells are a species of yeast that is an extremely popular expression cell line, which has produced some of the highest recombinant protein yields (*69*). This study showed the ability of MALDI-TOF to directly monitor fermentation broth for the production of a secreted AX2 protein (5 kDa) over time (*68*). At various time intervals, culture aliquots were centrifuged and the supernatant was analyzed by MALDI-TOF without further purification. One μL of MALDI matrix (α-cyano at 15 mg/mL in 70% ACN, 0.1% TFA) was spotted followed by one μL of culture supernatant. The resulting mass spectra were extremely clean (Figure 6), which is likely due to the fact that the protein was secreted and therefore naturally separated from other cellular components. The dynamics of the protein expression could be clearly tracked, and showed that significant protein expression occurred at 35 hours after induction (*68*).

A more extensive example of rapid detection of recombinant proteins from crude eukaryotic cell cultures was published in 2002 (*60*). This study monitored crude culture samples of *P. pastoris*, Sf21 insect cells and human embryonic kidney cells for recombinant protein expression via MALDI-TOF. *P. pastoris* cells were monitored for the expression of a secreted protein, serum amyloid P component (28.5 kDa), following methanol induction. An aliquot of crude cell culture media was concentrated and desalted via a reversed phase microcolumn (*70*), and spotted directly on the MALDI plate for analysis. Sinapinic acid was used as the MALDI matrix with the sandwich method. The resulting mass spectrum (Figure 7) shows a strong signal at the mass of this recombinant protein, likely due to the secretion of the protein and microcolumn cleanup. According to the authors, this method has now replaced an ELISA assay that required 2 days to perform. The authors also attempted to monitor the expression of this protein prior to secretion. The authors evaluated lithium acetate solution as well as 10% TFA to selectively solubilize the protein. Both methods were successful in producing spectra with a mass matching that of the recombinant protein. However, the spectra were much noisier and poorer quality than that of the secreted protein. Additionally, it is uncertain that the protein source was actually intracellular and not residual secreted protein being observed. Nonetheless, the detection of the secreted protein was extremely successful and replaced the conventional ELISA assay (*60*).

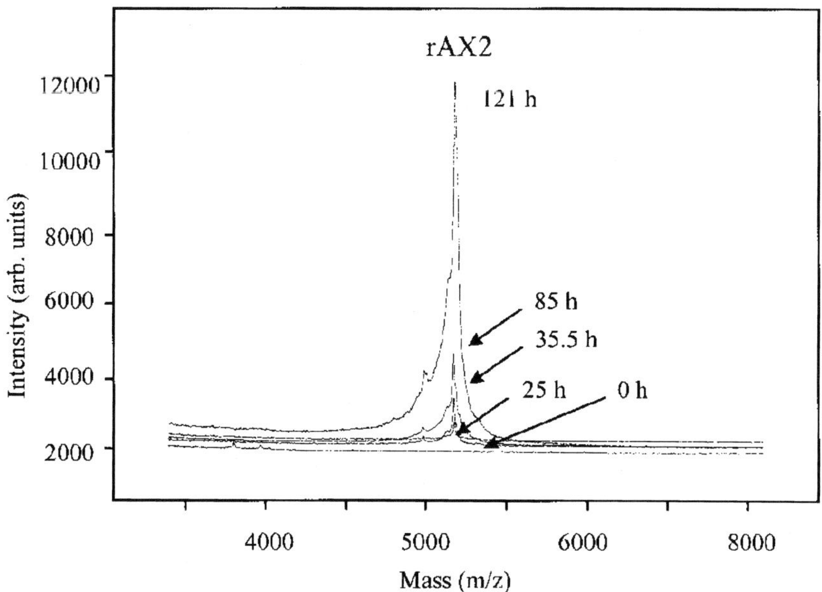

Figure 6. MALDI-TOF mass spectrum showing the production of AX2 recombinant protein expressed in Pichia pastoris at different time intervals. Reproduced with permission from reference (68). Copyright 1999, Academic Press.

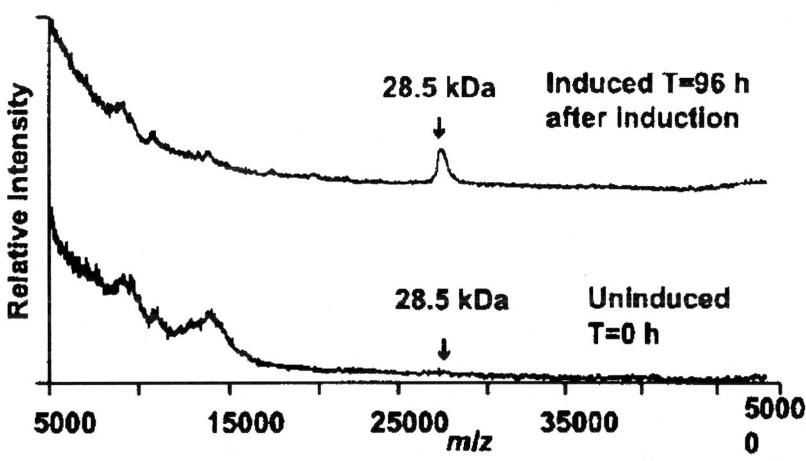

Figure 7. MALDI-TOF mass spectra of culture media from Pichia expressing a secreted recombinant protein (SAPC) at two time intervals; 0 hours and 96 hours after induction. Reproduced with permission from reference (60). Copyright 2002, Elsevier Science-USA.

This study also monitored the expression of plasmid-borne protein CrmD (36.8 kDa), a pox virus-encoded tumor necrosis factor receptor homologue expressed in insect cells (*60*, *71*). Baculovirus was used to infect Sf21 insect cells. The cells were harvested after 48 hours and spotted directly on a MALDI plate with sinapinic acid using the sandwich method. A peak at 36.7 kDa was clearly observable in the mass spectrum from cells expressing the CrmD recombinant protein, while the control cell line lacked a peak at this mass.

Bottom-Up/Proteomics Approaches

Viral Vectors in Prokaryotic Microorganisms

Gaining amino acid sequence information greatly increases the confidence in the identification of an unknown protein. To this end, a method demonstrating an on-probe digestion of a MS2 viral-borne protein directly from a viral suspension was published in 2002 (*72*). It should be pointed out that the crude culture with suspended virus was semipurified by centrifugation through a molecular weight cutoff filter (*72*). This solution was spotted on a MALDI plate and digested with trypsin for 20 minutes at room temperature. The MALDI source was mated to a FT-ICR mass analyzer, which allowed for tandem MS capability (*72*). A single peptide was observed at *m/z* 1753.957, which was isolated, and fragmented. The resulting tandem MS data was searched via MASCOT against the SwissProt database without taxonomic restriction. This search resulted in linking the observed peptide to coat proteins from three potential enterobacteriophage sources (MS2, R17, and F2), which are extremely similar genetically (*26*).

The characterization of this viral protein was further improved with the implementation of a residue-specific chemical cleavage (*73*). An aliquot of MS2 infected *E. coli* culture media was semipurified by centrifugation at 10,000 x g and passed through a 0.25 μm filter to remove cellular debris. Two hundred μL of this crude viral suspension was mixed with 200 μL of 50 % acetic acid and 50 μL of 1 % Triton X-100 detergent. This mixture was incubated in a CEM Discover Benchmate microwave system at 190 Watts for 30 seconds. This acid digestion cleaves with high specificity for aspartic acid residues (*74*, *75*). Figure 8 shows the resulting MALDI-TOF mass spectrum in which 100 % sequence coverage for the viral capsid protein was achieved. This result demonstrated the ability of MALDI-TOF to fully characterize the sequence of the viral protein from a crude bacterial extract (*73*). It should be mentioned that while Triton X-100 improved overall spectral quality, it was not required for complete coat protein denaturation and digestion (*73*).

Viral Vectors in Eukaryotic Cells

Recently, microwave assisted acid hydrolysis has been extended to the characterization of human adenovirus type 5 (*76*). The virus was grown in HeLa cells to the point of cytopathic lysis. As soon as lysis was observed, the HeLa cells were collected by centrifugation (8,000 x g for 20 min) washed twice with 0.1M PBS (pH = 7.1) and resuspended in 10mM Tris-HCl buffer. This crude

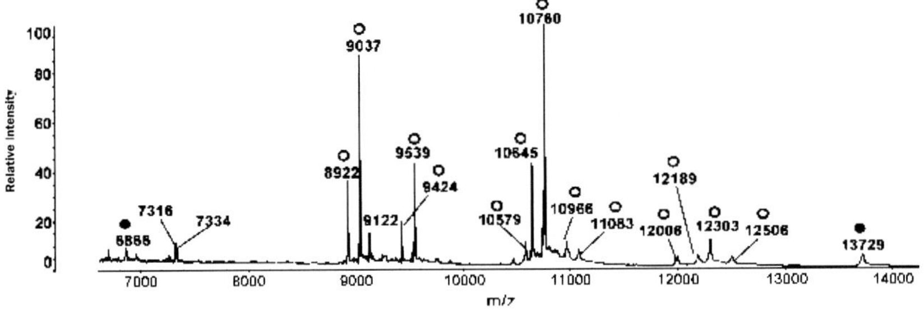

Figure 8. MALDI-TOF mass spectrum of bacteriophage MS2 suspension following a microwave-assisted acid digestion for 30 seconds. (○ = acid digestion products, ● = undigested protein). Reproduced with permission from reference (73). Copyright 2007, American Chemical Society.

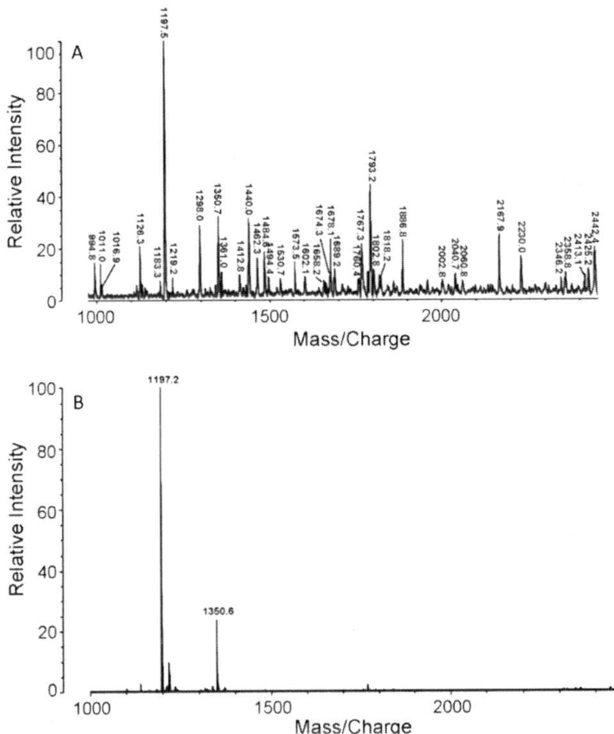

Figure 9. MALDI-TOF mass spectra of peptides from human adenovirus type 5 (a) following a 2 minute acid digestion, and (b) no digestion performed. Reproduced with permission from reference (76). Copyright 2010, Elsevier B.V.

sample was subjected to microwave assisted acid hydrolysis via incubation in 12.5 % acetic acid at 140 °C for 2 minutes. The resulting digestion products were spotted on a sample plate for MALDI-TOF analysis. The resulting MALDI-TOF spectrum is shown in Figure 9a, along with an undigested control spectrum in Figure 9b. A total of 37 peptides were identified in the digested sample and linked to 8 adenovirus proteins. The two peptides found in the undigested virus sample were linked to endogenous viral protease cleavage products. The authors also attempted this analysis via a trypsin digestion and reported that few tryptic peptides were observed. This comparison illustrates the advantage of the acid digestion in which the virus proteins are denatured in the acidic digestion solution making them more accessible for cleavage. The authors also successfully demonstrated the use of O^{18} labeled water in the digestion solution to quantify the growth of the adenovirus in the HeLa cells over time (*76*).

Plasmid Vectors in Prokaryotic Microorganisms

Bottom up proteomics strategies have also been applied to plasmid-borne proteins. *E. coli* cells that have been transformed with a plasmid often require selective pressure to preserve those cells that harbor the plasmid (*1*, *6*). Antibiotic resistance is often encoded by the plasmid, allowing for growth in antibiotic enriched media. The most commonly used antibiotic resistance gene is "ampR" or "bla", which codes for β-lactamase (*77*). β-lactamase confers resistance to the penicillin class of antibiotics by cleaving their lactam ring and rendering them ineffective (*78*, *79*). It is often important to ensure that the cells have retained the plasmid prior to induction. MALDI-TOF was evaluated for this application in 2007 (*62*). In order to quickly confirm plasmid insertion by whole cell MALDI-TOF, an *E. coli* cell suspension (2mg/mL) was spotted with sinapinic acid via the sandwich method (*62*). The presence of the plasmid was confirmed by the observation of a peak at *m/z* 28,907, which corresponds to that of β-lactamase (28,908 Da) shown in Figure 10.

To further confirm the identity of this plasmid-borne protein, a 20 minute on-probe tryptic digestion was performed on the crude cell suspension. The resulting MALDI-TOF mass spectrum is shown in Figure 11. The observed tryptic peptides were fragmented and the tandem MS data was searched via MASCOT against all entries in the NCBInr database without taxonomic restriction. This search resulted in a definitive identification of β-lactamase with a MASCOT score of 127, well above the 99 % confidence threshold score of 58. Additionally, host *E. coli* microorganism peptides were identified (*62*). This method would be amenable to quickly validating insertion of plasmids that contain the "ampR" or "bla" gene, which is extremely common and found in ~80% of cloning vectors annotated by ATCC (*62*, *80*).

This work was repeated independently by another group in which *E. coli* cells were transformed with an ampR containing plasmid (*63*). The cells were subjected to a more elaborate and time consuming extraction protocol, in which the cells were centrifuged at 14,000 x g for 5 minutes at 4°C. The cell pellets were then washed twice with 0.1% TFA and resuspended in 100 μL of extraction solvent (formic acid: isopropyl alcohol: water at 17:33:50 ratio by volume). This suspension was vortexed for one minute, and centrifuged at 14,000 x g for 2 minutes at 24°C. The supernatant was then spotted on the MALDI plate and mass analyzed by time-of-flight. Cells transformed with the ampR containing plasmid produced a signal at ~28.9 kDa matching that of β-lactamase. Interestingly, β-lactamase was observed in *E. coli* that were transformed with the plasmid, but not

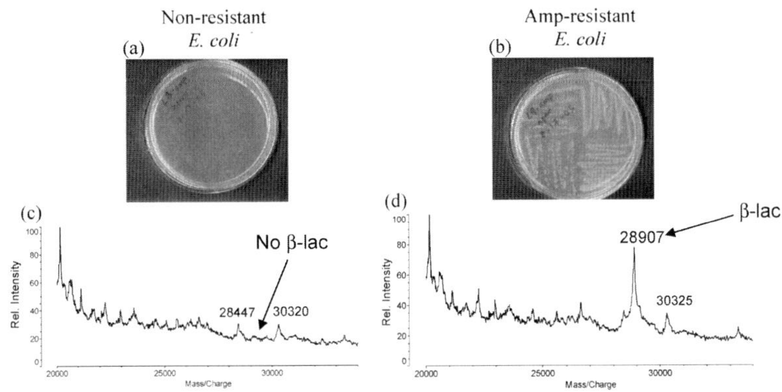

Figure 10. Images of E. coli cells (a) without & (b) with plasmids grown overnight on LB-agar plates containing ampicillin (100μg/mL). MALDI-TOF mass spectra from suspensions of (c) nonresistant E. coli cells and (d) ampicillin resistant E. coli cells. Reproduced with permission from reference (62). Copyright 2007, American Chemical Society.

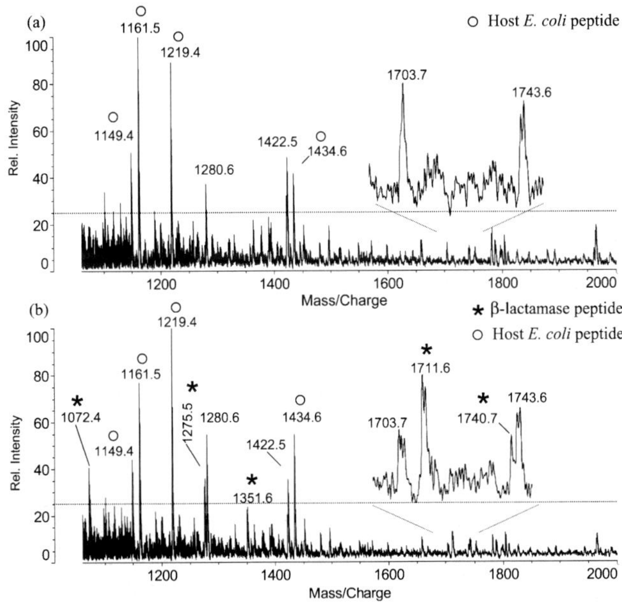

Figure 11. MALDI-TOF mass spectra from suspensions of (a) E. coli control and (b) plasmid containing E. coli. Both were subjected to on-probe trypsin digestions. Reproduced with permission from reference (62). Copyright 2007, American Chemical Society.

kept under selective pressure. This result was contrary to what was observed in the previous study by Russell *et. al* (*62*). These conflicting results could reflect a difference in the degree and rate with which each plasmid is rejected by the host *E. coli* cells (*6*).

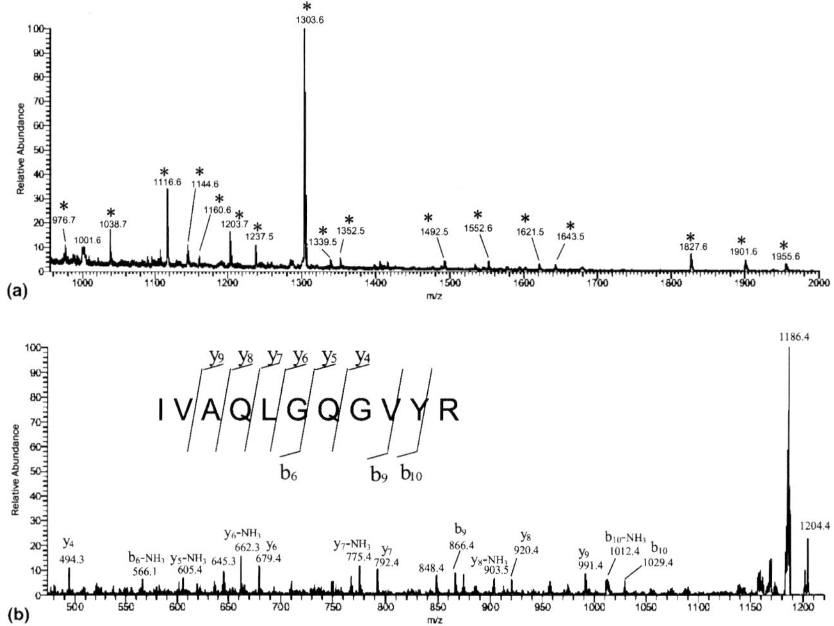

Figure 12. (a) AP-MALDI mass spectrum of peptide products resulting from a basic solubilization of δ-endotoxin proteins directly from B. thuringiensis spores followed by a 5-min trypsin digestion. Tryptic peptide mass matches to those from δ-endotoxins are indicated with asterisks. (b) Tandem mass spectrum of m/z 1203.7 yielding peptide sequence information. Reproduced with permission from reference (81). Copyright 2010, American Society for Mass Spectrometry published by Elsevier B.V.

In a 2010 paper, a native plasmid-borne protein was detected and identified via whole cell AP-MALDI performed on *Bacillus thuringiensis* var. kurstaki HD-1 spores (*81*). AP-MALDI offers the advantage of reduced in-source fragmentation compared with conventional MALDI, which can reduce spectral complexity (*82–85*). The AP-MALDI ion source was mated to a linear ion trap mass analyzer, allowing for tandem MS capability (*81*). *B. thuringiensis* is not harmful to humans, but targets insects making it a popular natural insecticide (*86*). *B. thuringiensis* produces high molecular weight δ-endotoxins (*87*). These pore forming toxins are soluble under alkaline conditions found in the midgut of many insects, and cause the rupture of the insects digestive system (*87*). These plasmid-borne proteins are known to be overexpressed and secreted in crystalline form when *B. thuringiensis* is in the spore form (*88*). In order to quickly detect and identify these proteins, the authors suspended the spores (10 mg/mL) in 100 mM potassium hydroxide (pH = 13). The suspension was vortexed for one minute, and allowed to sit for an additional 19 minutes to ensure complete δ-endotoxin solubilization (*87*, *89*). The pH was then adjusted to 8 by dropwise addition of 1M HCl. A 100 μL aliquot of this crude extraction mixture was combined with an equal volume of a 50 % by mass suspension of immobilized trypsin and allowed to digest for 5 minutes at room temperature (*39*, *52*). The immobilized trypsin offers the advantage of easy removal from the digestion mixture by centrifugation. Additionally, immobilized trypsin digests lack trypsin autolysis peaks (*90*). It was estimated that ~ 50 μg of total δ-endotoxin was present in the digestion mixture, based on known δ-endotoxin expression levels (*91*, *92*). The digestions were quenched via centrifugation at 10,000 x g for 3 minutes to remove spores and immobilized trypsin. A 10 μL aliquot of the supernatant was cleaned up and concentrated to 5 μL via a μC-18 Ziptip, and spotted on the AP-MALDI plate for MS analysis.

δ-endotoxin tryptic peptides that were known to be unique to each toxin based on MS-BLAST searches (*93*) were targeted for tandem MS. Figure 12a shows a mass spectrum of the crude mixture following a 5 minute digestion (*81*). Seventeen δ-endotoxin tryptic peptides were observed, and nine of which produced tandem mass spectra. Figure 12b shows a tandem mass spectrum of *m/z* 1203.7, yielding peptide sequence information.

The tandem MS data was pooled and searched against all entries in the SwissProt/Trembl database using the MASCOT MS/MS ion search engine without taxonomic restriction (*94*). The search resulted in a definitive protein match to the δ-endotoxin Cry1Ab with a MASCOT score of 278 and expect value of 7.5×10^{-23}. This protein has a molecular weight of ~133 kDa and demonstrates the advantage of coupling selective solubilization with a bottom up approach to quickly detect and identify a high mass plasmid-borne protein from a crude sample (*81*).

Conclusions

The methods summarized in this chapter demonstrate the ability of MALDI-MS to quickly detect recombinant proteins directly from crude cell cultures. This has been shown to be possible for both viral and plasmid vectors in both prokaryotic and eukaryotic expression systems. Rapid recombinant protein identification with high confidence has also been shown to be possible by utilizing enzymatic or chemical digestions to yield sequence information. The use of MALDI-MS for high throughput screening of cultures for recombinant protein expression opens up the possibility of improved protein expression by optimizing expression conditions. This level of optimization is not possible with conventional methods for recombinant protein detection due to their lengthy protocols. Future efforts may focus on top-down efforts to characterize recombinant protein expression with high confidence without the need for enzymatic or chemical digestions.

References

1. Glick, B. R.; Pasternak, J. J. *Molecular Biotechnology: Principles and Applications of Recombinant DNA*; 3rd ed.; ASM Press: Washington, DC, 2003.

2. Itakura, K.; Hirose, T.; Crea, R.; Riggs, A. D.; Heyneker, H. L.; Bolivar, F.; Boyer, H. W. *Science* **1977**, *198* (4321), 1056–1063.
3. Goeddel, D. V.; Kleid, D. G.; Bolivar, F.; Heyneker, H. L.; Yansura, D. G.; Crea, R.; Hirose, T.; Kraszewski, A.; Itakura, K.; Riggs, A. D. *Proc. Natl. Acad. Sci. U.S.A.* **1979**, *76* (1), 106–110.
4. Swartz, J. R. *Curr. Opin. Biotechnol.* **2001**, *12* (2), 195–201.
5. Makrides, S. C. *Microbiol. Rev.* **1996**, 512–538.
6. Baneyx, F. *Curr. Opin. Biotechnol.* **1999**, *10* (5), 411–421.
7. Walsh, G.; Jefferis, R. *Nat. Biotechnol.* **2006**, *24* (10), 1241–1252.
8. Hanes, B. D.; Rickwood, D. *Gel Electrophoresis of Proteins: A Practical Approach*; Oxford University Press: New York, 1990.
9. O'Farrell, P. H. *J. Biol. Chem.* **1975**, *250*, 4007–4021.
10. Strahler, J. R.; Kuick, R.; Hanash, S. M. In *Protein Structure: A practical approach*; Creighton, T., Ed.; IRL Press: Oxford, U.K., 1989; pp 231−266.
11. Henzel, W. J.; Billeci, T. M.; Stults, J. T.; Wong, S. C.; Grimley, C.; Watanabe, C. *Proc. Natl. Acad. Sci. U.S.A.* **1993**, *90* (11), 5011–5015.
12. Liang, X.; Bai, J.; Liu, Y.; Lubman, D. M. *Anal. Chem.* **1996**, *68* (6), 1012–1018.
13. Shevchenko, A.; Wilm, M.; Vorm, O.; Mann, M. *Anal. Chem.* **1996**, *68* (5), 850–858.
14. Courchesne, P. L.; Luethy, R.; Patterson, S. D. *Electrophoresis* **1997**, *18* (3−4), 369–381.
15. O'Connell, K. L.; Stults, J. T. *Electrophoresis* **1997**, *18* (3−4), 349–359.
16. Schuhmacher, M.; Glocker, M. O.; Wunderlin, M.; Przybylski, M. *Electrophoresis* **1996**, *17* (5), 848–854.
17. Ogorzalek Loo, R. R.; Stevenson, T.; Mitchell, C.; Loo, J. A.; Andrews, P. C. *Anal. Chem.* **1996**, *68* (11), 1910–1917.
18. Cohen, S. L.; Chait, B. T. *Anal. Biochem.* **1997**, *247* (2), 257–267.
19. Anhalt, J. P.; Fenselau, C. *Anal. Chem.* **1975**, *47* (2), 219–225.
20. Snyder, P. A.; Smith, P. B. W.; Dworzanski, J. P.; Meuzelaar, H. L. C. Pyrolysis gas chromatography mass spectrometry. In *Mass Spectrometry for the Characterization of Microorganisms*; American Chemical Society: Washington, DC, 1993; Vol. 541; pp 62−84.
21. Basile, F.; Beverly, M. B.; Voorhees, K. J.; Hadfield, T. L. *Trends Anal. Chem.* **1998**, *17*, 95–109.
22. Heller, D. N.; Cotter, R. J.; Fenselau, C.; Uy, O. M. *Anal. Chem.* **1987**, *59* (23), 2806–2809.
23. Despeyroux, D.; Phillpotts, R.; Watts, P. *Rapid Commun. Mass Spectrom.* **1996**, *10* (8), 937–941.
24. Liu, C.; Hofstadler, S. A.; Bresson, J. A.; Udseth, H. R.; Tsukuda, T.; Smith, R. D.; Snyder, P. *Anal. Chem.* **1998**, *70* (9), 1797–1801.
25. Xiang, F.; Anderson, G. A.; Veenstra, T. D.; Lipton, M. S.; Smith, R. D. *Anal. Chem.* **2000**, *72* (11), 2475–2481.
26. Cargile, B. J.; McLuckey, S. A.; Stephenson, J. L. *Anal. Chem.* **2001**, *73* (6), 1277–1285.
27. Claydon, M. A.; Davey, S. N.; Edwards-Jones, V.; Gordon, D. B. *Nat. Biotechnol.* **1996**, *14*, 1584–1586.

28. Holland, R. D.; Wilkes, J. G.; Rafii, F.; Sutherland, J. B.; Persons, C. C.; Voorhees, K. J.; Lay, J. O. *Rapid Commun. Mass Spectrom.* **1996**, *10* (10), 1227–1232.
29. Krishnamurthy, T.; Rajamani, U.; Ross, P. L. *Rapid Commun. Mass Spectrom.* **1996**, *10* (8), 883–888.
30. Fenselau, C.; Demirev, P. A. *Mass Spectrom. Rev.* **2001**, *20* (4), 157–171.
31. Thomas, J. J.; Falk, B.; Fenselau, C.; Jackman, J.; Ezzell, J. *Anal. Chem.* **1998**, *70* (18), 3863–3867.
32. Karty, J. A.; Susan, L., S.; Reilly, J. P. *Rapid Commun. Mass Spectrom.* **1998**, *12* (10), 625–629.
33. Scholl, P. F.; Leonardo, M. A.; Rule, A. M.; Carlson, M. A.; Antoine, M. D.; Buckley, T. J. *Johns Hopkins APL Tech. Dig.* **1999**, *20*, 343–351.
34. Ryzhov, V.; Hathout, Y.; Fenselau, C. *Appl. Environ. Microbiol.* **2000**, *66* (9), 3828–3834.
35. Wang, Z.; Russon, L.; Li, L.; Roser, D. C.; Long, S. R. *Rapid Commun. Mass Spectrom.* **1998**, *12* (8), 456–464.
36. Madonna, A. J.; Basile, F.; Ferrer, I.; Meetani, M. A.; Rees, J. C.; Voorhees, K. J. *Rapid Commun. Mass Spectrom.* **2000**, *14* (23), 2220–2229.
37. Winkler, M. A.; Uher, J.; Cepa, S. *Anal. Chem.* **1999**, *71* (16), 3416–3419.
38. Evason, D. J.; Claydon, M. A.; Gordon, D. B. *Rapid Commun. Mass Spectrom.* **2000**, *14* (8), 669–672.
39. Warscheid, B.; Fenselau, C. *Anal. Chem.* **2003**, *75* (20), 5618–5627.
40. Arnold, R. J.; Reilly, J. P. *Anal. Biochem.* **1999**, *269* (1), 105–112.
41. Dai, Y.; Li, L.; Roser, D.; Long, S. R. *Rapid Commun. Mass Spectrom.* **1999**, *13* (1), 73–78.
42. Holland, R. D.; Duffy, C. R.; Rafii, F.; Sutherland, J. B.; Heinze, T. M.; Holder, C. L.; Voorhees, K. J.; Lay, J. O., Jr. *Anal. Chem.* **1999**, *71* (15), 3226–3230.
43. Demirev, P. A.; Ramirez, J.; Fenselau, C. *Anal. Chem.* **2001**, *73* (23), 5725–5731.
44. Haag, A. M.; Taylor, S. N.; Johnston, K. H.; Cole, R. B. *J. Mass Spectrom.* **1998**, *33* (8), 750–756.
45. Saenz, A. J.; Petersen, C. E.; Valentine, N. B.; Gantt, S. L.; Jarman, K. H.; Kingsley, M. T.; Wahl, K. L. *Rapid Commun. Mass Spectrom.* **1999**, *13* (15), 1580–1585.
46. Jarman, K. H.; Cebula, S. T.; Saenz, A. J.; Petersen, C. E.; Valentine, N. B.; Kingsley, M. T.; Wahl, K. L. *Anal. Chem.* **2000**, *72* (6), 1217–1223.
47. Conway, G. C.; Smole, S. C.; Sarracino, D. A.; Arbeit, R. D.; Leopold, P. E. *J. Mol. Microbiol. Biotechnol.* **2001**, *1*, 103–112.
48. Demirev, P. A.; Ho, Y.; Ryzhov, V.; Fenselau, C. *Anal. Chem.* **1999**, *71* (14), 2732–2738.
49. Demirev, P. A.; Lin, J. S.; Pineda, F. J.; Fenselau, C. *Anal. Chem.* **2001**, *73* (19), 4566–4573.
50. Pineda, F. J.; Lin, J. S.; Fenselau, C.; Demirev, P. A. *Anal. Chem.* **2000**, *72* (16), 3739–3744.
51. Yao, Z.; Demirev, P. A.; Fenselau, C. *Anal. Chem.* **2002**, *74* (11), 2529–2534.

52. Warscheid, B.; Jackson, K.; Sutton, C.; Fenselau, C. *Anal. Chem.* **2003**, *75* (20), 5608–5617.
53. Demirev, P. A.; Feldman, A. B.; Kowalski, P.; Lin, J. S. *Anal. Chem.* **2005**, *77* (22), 7455–7461.
54. Fagerquist, C. K.; Garbus, B. R.; Miller, W. G.; Williams, K. E.; Yee, E.; Bates, A. H.; Boyle, S.; Harden, L. A.; Cooley, M. B.; Mandrell, R. E. *Anal. Chem.* **2010**, *82* (7), 2717–2725.
55. Teresa, C. C.; David, M. L.; Walter, J. W.; Vertes, A. *Rapid Commun. Mass Spectrom.* **1994**, *8* (12), 1026–1030.
56. Parker, C. E.; Papac, D. I.; Tomer, K. B. *Anal. Biochem.* **1996**, *239* (1), 25–34.
57. Easterling, M. L.; Colangelo, C. M.; Scott, R. A.; Amster, J. I. *Anal. Chem.* **1998**, *70* (13), 2704–2709.
58. Winkler, M. A.; Hickman, R. K.; Golden, A.; Aboleneen, H. *Biotechniques* **2000**, *28*, 890–892, 894–895.
59. Villanueva, J.; Canals, F.; Querol, E.; Avilés, F. *Enzyme Microb. Technol.* **2001**, *29* (1), 99–103.
60. Jebanathirajah, J. A.; Andersen, S.; Blagoev, B.; Roepstorff, P. *Anal. Biochem.* **2002**, *305* (2), 242–250.
61. Kulkarni, M. J.; Vinod, V. P.; Umasankar, P. K.; Patole, M. S.; Rao, M. *Rapid Commun. Mass Spectrom.* **2006**, *20* (18), 2769–2772.
62. Russell, S. C.; Edwards, N.; Fenselau, C. *Anal. Chem.* **2007**, *79* (14), 5399–5406.
63. Camara, J.; Hays, F. *Anal. Bioanal. Chem.* **2007**, *389* (5), 1633–1638.
64. Dai, Y.; Whittal, R. M.; Li, L. *Anal. Chem.* **1999**, *71* (5), 1087–1091.
65. Jones, J.; Wilkins, C. L.; Cai, Y.; Beitle, R. R.; Liyanage, R.; Jackson, O. L. *Biotechnol. Prog.* **2005**, *21* (6), 1754–1758.
66. Jones, J. J.; Stump, M. J.; Fleming, R. C.; Lay, J. O.; Wilkins, C. L. *Anal. Chem.* **2003**, *75* (6), 1340–1347.
67. Valegard, K.; Liljas, L.; Fridborg, K.; Unge, T. *Nature* **1990**, *345* (6270), 36–41.
68. Kristensen, A. K.; Brunstedt, J.; Nielsen, J. E.; Mikkelsen, J. D.; Roepstorff, P.; Nielsen, K. K. *Protein Expression Purif.* **1999**, *16* (3), 377–387.
69. Smith, C. *The Scientist* **1998**, *12* (22), 20–28.
70. Kussmann, M.; Roepstorff, P. Sample Preparation Techniques for Peptides and Proteins Analyzed by MALDI-MS. *Mass Spectrometry of Proteins and Peptides*; Springer: 2000; pp 405−424.
71. Saraiva, M.; Alcami, A. *J. Virol.* **2001**, *75* (1), 226–233.
72. Yao, Z.; Afonso, C.; Fenselau, C. *Rapid Commun. Mass Spectrom.* **2002**, *16* (20), 1953–1956.
73. Swatkoski, S.; Russell, S.; Edwards, N.; Fenselau, C. *Anal. Chem.* **2007**, *79* (2), 654–658.
74. Inglis, A. S. Cleavage at aspartic acid. *Methods in Enzymology*; Academic Press: 1983; Vol. 91, pp 324−332.
75. Li, A.; Sowder, R. C.; Henderson, L. E.; Moore, S. P.; Garfinkel, D. J.; Fisher, R. J. *Anal. Chem.* **2001**, *73* (22), 5395–5402.

76. Fenselau, C.; Laine, O.; Swatkoski, S. *Int. J. Mass Spectrom.* **2010**, in press.
77. Jana, S.; Deb, J. K. *Appl. Microbiol. Biotechnol.* **2005**, *67* (3), 289–298.
78. Koch, A. L. *Crit. Rev. Microbiol.* **2000**, *26* (4), 205–220.
79. Walsh, C. *Nature* **2000**, *406* (6797), 775–781.
80. http://www.atcc.org.
81. Nguyen, J.; Russell, S. C. *J. Am. Soc. Mass Spectrom.* **2010**, *21* (6), 993–1001.
82. Doroshenko, V. M.; Laiko, V. V.; Taranenko, N. I.; Berkout, V. D.; Lee, H. S. *Int. J. Mass Spectrom.* **2002**, *221* (1), 39–58.
83. Madonna, A. J.; Voorhees, K. J.; Taranenko, N. I.; Laiko, V. V.; Doroshenko, V. M. *Anal. Chem.* **2003**, *75* (7), 1628–1637.
84. Laiko, V. V.; Moyer, S. C.; Cotter, R. J. *Anal. Chem.* **2000**, *72* (21), 5239–5243.
85. Laiko, V. V.; Baldwin, M. A.; Burlingame, A. L. *Anal. Chem.* **2000**, *72* (4), 652–657.
86. Eizaguirre, M.; Tort, S.; Lopez, C.; Albajes, R. *J. Econ. Entomol.* **2005**, *98* (2), 464–470.
87. Lee, K.; Kang, E.; Park, S.; Ahn, S.; Yoo, K.; Kim, J.; Lee, H. *Proteomics* **2006**, *6* (5), 1512–1517.
88. Bechtel, D. B.; Bulla, L. A. *J. Bacteriol.* **1976**, *127*, 1472–1481.
89. Tyrell, D. J.; Bulla, L. A.; Andrews, R. E.; Kramer, K. J.; Davidson, L. I.; Nordin, P. *J. Bacteriol.* **1981**, *145*, 1052–1062.
90. Vestling, M.; Murphy, C.; Fenselau, C. *Anal. Chem.* **1990**, *62* (21), 2391–2394.
91. Agaisse, H.; Lereclus, D. *J. Bacteriol.* **1995**, *177*, 6027–6032.
92. Andrews, R. E.; Iandolo, J. J.; Campbell, B. S.; Davidson, L. I.; Bulla, L. A. *Appl. Environ. Microbiol.* **1980**, *40*, 897–900.
93. Shevchenko, A.; Sunyaev, S.; Loboda, A.; Shevchenko, A.; Bork, P.; Ens, W.; Standing, K. G. *Anal. Chem.* **2001**, *73* (9), 1917–1926.
94. Perkins, D. N.; Pappin, D. J. C.; Creasy, D. M.; Cottrell, J. S. *Electrophoresis* **1999**, *20* (18), 3551–3567.

Chapter 6

Detection, Differentiation and Subtyping of Botulinum Neurotoxins in Clinical Samples with Mass Spectrometry

John R. Barr,* Suzanne R. Kalb, and James L. Pirkle

Centers for Disease Control and Prevention, National Center for Environmental Health, Division of Laboratory Sciences, 4770 Buford Hwy, N.E., Atlanta, GA 30341
***To whom correspondence should be addressed. John R. Barr, Ph.D.: e-mail jbarr@cdc.gov.**

Botulinum neurotoxins (BoNTs) are the most toxic substances known to man and are the etiologic agents responsible for the deadly disease botulism. To provide a means of assessing exposure to BoNTs, we have developed the Endopep-MS method, which can detect and differentiate BoNTs in clinical and food samples. The method involves extracting the toxin with high affinity antibodies that are selective for only one of the BoNT serotypes; cleavage of synthetic peptides according to the specific enzymatic activity for each of the BoNT serotypes; and selectively detecting the toxin-dependent peptides by using high resolution matrix-assisted laser–desorption ionization time-of-flight mass spectrometry. We can achieve limits of detection lower than that of the mouse bioassay, the historic standard, in clinical samples with BoNT/A, /B, /E, and /F, the four serotypes typically associated with human cases of botulism. We discuss the application of the method with two examples: an outbreak of botulism from hot dog chili sauce and botulism in Mississipi catfish. We have also added proteomic analysis to identify BoNT subtypes and strains allowing epidemiologists to quickly assess possible similarities and differences between concurrent botulism outbreaks.

Introduction

Botulinum neurotoxins (BoNT) are the most toxic substances known (*1*). They are produced under anaerobic conditions by strains of *Clostridium botulinum*, *C. butyricum*, *C. baratii*, and *C. argentinese* (*1*, *2*). Intoxication with one of the seven distinct serotypes of BoNT (A-G) causes the deadly disease known as botulism, which is characterized by a flaccid paralysis (*2*). Four serotypes of BoNT, /A, /B, /E, and /F, are known to cause botulism in humans (*2*). Typically, human cases of botulism are from ingestion of foods that contain the toxins or colonization of toxin producing species of *Clostridia* in the gastrointestinal tracts of infants or immunocompromised adults or in wounds (*2*). In 2008, the CDC reported 153 cases of botulism in the US (*3*). The 2008 botulism cases included 12% foodborne, 73% infant, 15% wound and 1% of unknown etiology. The 18 foodborne botulism cases reported in 2008 were caused by type A toxin (56%) and type E toxin (33%) with 11% unknown toxin type. The infant botulism cases were caused by type A toxin (45%), type B toxin (54%), type F toxin (1%) and a dual toxin producing (Bf) *C. botulinum* species (1%). The 15 wound botulism cases were all from type A toxin (*3*).

BoNTs are zinc metalloproteases that cleave and inactivate specific cellular proteins that are essential for the release of the neurotransmitter acetylcholine (Figure 1). Each of the BoNT toxins are composed of a heavy chain which is about 100,000 daltons and a light chain of about 50,000 daltons (*1*). The heavy chain is responsible for receptor binding and delivering the catalytic light chain into the area of target neurons (*4*, *5*). The light chain is the zinc metalloprotease portion of the toxin that selectively cleaves neuronal proteins required for normal exocytosis (*1*). Although the light chain is responsible for the specific toxicity, the heavy chain is required to produce the *in vivo* toxicity.

Each BoNT serotype has a specific toxin dependent cleavage site. BoNT A, C and E cleave SNAP-25. BoNT A cleaves SNAP-25 between glutamine-196 and arginine-197 (*6–9*), BoNT C cleaves SNAP-25 at the adjacent residue between arginine-197 and alanine-198 (*10*, *11*), while BoNT E cleaves SNAP-25 between arginine-180 and isoleucine-181 (*7–9*). BoNT B, D, F, G all cleave synaptobrevin 2 (also called VAMP 2). The cleavage of synaptobrevin 2 by BoNT B, D, F, and G are at glutamaine-75 (*12*), lysine-58 (*8*, *13*), glutamine-57 (*13*), and alanine-61 (*14*, *15*) respectively. Of the serotypes mentioned, only BoNT C cleaves more than one site on a specific protein. In addition to cleaving SNAP-25, BoNT C also cleaves syntaxin between lysine-253 and alanine-254 (*16*, *17*).

In addition to the 7 serotypes of BoNT there are subtypes that at have similar serological properties. Within serotype A there are 5 known subtypes, A1-A5 (*18*, *19*). These currently known subtypes have between 2% to 16% differences in amino acid sequence. Within a subtype, there are different strains that can differ by as little as one amino acid difference. BoNT/B is currently divided into B1, B2, B3, nonproteolytic (np) B (B4), bivalent (bv) B (B5), and B6 subtypes (*18*, *20*), with an amino acid variance of 7% or less. E1, E2, E3, E4 (Italian *butyricum*), E5 (Chinese *butyricum*), and E6 subtypes currently comprise BoNT/E (*18*, *21*, *22*). Two E subtypes were isolated from BoNT E-producing *C. butyricum* strains. BoNT/F is divided into proteolytic F, npF, bvF and BoNT F-producing C. *baratii* subtypes (*18*). The BoNT E subtypes exhibit 5% or less amino acid variance while the known F subtypes have up to 32% variance.

It should be noted that botulinum neurotoxins are very toxic and must be handled using care and appropriate safety measures. All neurotoxins require handling in a level 2 biosafety cabinet equipped with HEPA filters.

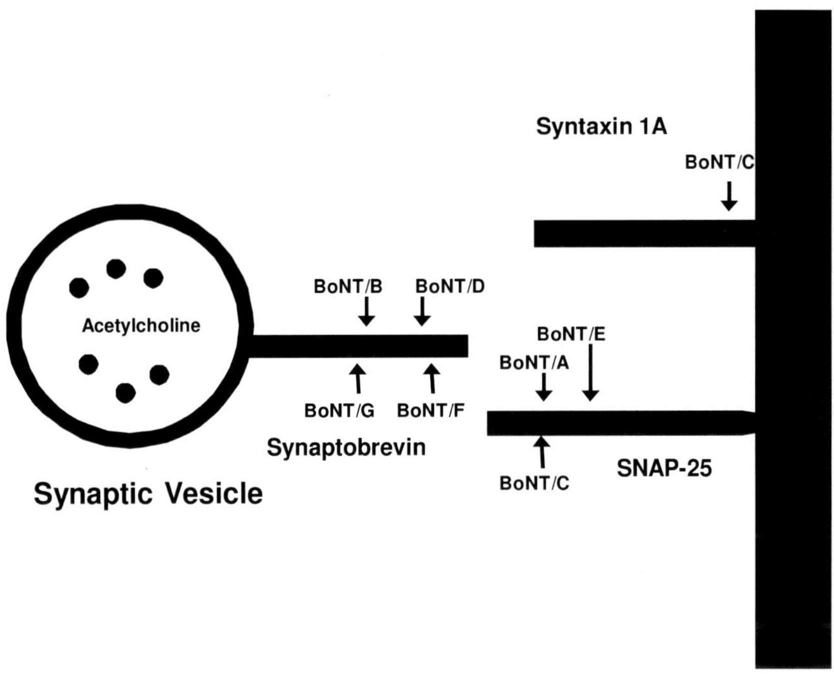

Figure 1. Schematic of the soluble N-ethylmaleimide sensitive fusion protein attachment receptor (SNARE) complex. All proteins must be intact for acetylcholine release. The cleavage sites of the proteins by BoNT are depicted.

Detection and Serotype Differentiation of Botulinum Neurotoxins

Historically, the mouse bioassay has been the most commonly used and accepted method to detect BoNT and confirm a diagnosis of botulism (*2*, *23*). The mouse bioassay employs mixtures of neutralizing antibodies given to mice in conjunction with the sample in question to differentiate the toxin serotype. Mice receiving the appropriate anti-BoNT serotype antibody with the toxic sample survive, while mice treated with the other serotype antibodies do not survive. The mouse bioassay is very sensitive detecting 1 mouse LD_{50}, which is thought to contain approximately 10 pg of active toxin for BoNT/A (*24*). However, the mouse bioassay is slow (taking up to 4 days) for final results, requires several grams of sample and requires a great deal of live animal use. Detection limits

Botulinum Neurotoxin Specific Peptide Substrates

	Sequence (cleavage by BoNT-A, BoNT-C)	m/z
1. BoNT-A, -C Substrate	Biotin-KGSNRTRIDQGNQRATRLLGGK-Biotin	2878.7
BoNT A cleavage site	Biotin-KGSNRTRIDQGNQ	1699.9
	RATRLLGGK-Biotin	1197.8
BoNT C cleavage site	Biotin-KGSNRTRIDQGNQR	1855.9
	ATRLLGGK-Biotin	1041.8
	BoNT-B BoNT-G	
2. BoNT-B, -G Substrate	LSELDDRADALQAGASQFESSAAKLKRKYWWKNLK	4023.8
BoNT B cleavage site	LSELDDRADALQAGASQ	1759.7
	FESSAAKLKRKYWWKNLK	2283.1
BoNT G cleavage site	LSELDDRADALQAGASQFESSA	2280.7
	AKLKRKYWWKNLK	1762.1
	BoNT-F BoNT-D	
3. BoNT-D, -F Substrate	LQQTQAQVDEVVDIMRVNVDKVLERDQKLSELDDRADAL	4497.1
BoNT F cleavage site	LQQTQAQVDEVVDIMRVNVDKVLERDQ	3168.9
	KLSELDDRADAL	1345.8
BoNT D cleavage site	LQQTQAQVDEVVDIMRVNVDKVLERDQK	3296.9
	LSELDDRADAL	1217.8
	BoNT-E	
4. BoNT-ESubstrate	IIGNLRHMALDMGNEIDTQNRQIDRIMEKADSNKT	4041.1
BoNT E cleavage site	IIGNLRHMALDMGNEIDTQNRQIDR	2923.4
	IMEKADSNKT	1136.5

Figure 2. Amino acid sequences of peptide substrates used in the Endopep-MS assay for detection and differentiation of botulinum neurotoxins (BoNT). Cleavage products of each serotype of BoNT and their observed m/z are also present. Peptide 1 is a mimic of the natural substrate for BoNT/A and /C whose natural target is SNAP-25. Peptide 2 is the substrate for BoNT/B and /G and is a mimic for VAMP 2. Peptide 3 is the substrate for BoNT/D and /F and is a mimic for VAMP-2, and peptide 4 is the substrate for BoNT/E and is a mimic for SNAP-25.

for *in vitro* methods and purified BoNT standards are generally reported as the number of mouse LD_{50} and therefore in this chapter, we will hold to the same convention.

To improve the detection of BoNT, a mass spectrometry based method called Endopep-MS has been developed (*25–30*). The Endopep-MS method detects BoNT and determines the serotype at the same time. It is similar in concept to previous non-mass spectrometry based *in vitro* methods for BoNT/A and /B (*31–34*) . The Endopep-MS method is based on extraction with antibodies specific to the toxin serotype (*27*). A reaction of the toxin with synthetic peptides that mimic the natural targets of the toxin produces unique cleavage products that can be differentiated by mass spectrometry (Figure 2) (*25–30*). Since each BoNT serotype has a different cleavage site, the product peptides not only detect the toxin activity but also differentiate the serotype.

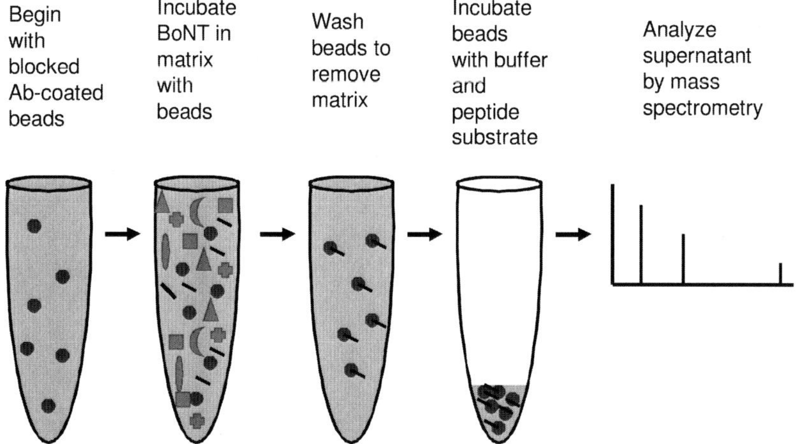

Figure 3. Protocol for sample preparation for Endopep-MS method to detect and differentiate botulinum neurotoxins (BoNT). Magnetic beads are coated with serotype specific antibodies and then incubated with the sample to extract BoNT. The beads are washed to remove nonspecific binding compounds and then incubated with peptide mimics to the natural substrate of the BoNT. The cleavage of the peptide mimic is then detected by matrix assisted laser desorption mass spectrometry.

Extraction of BoNT from Clinical, Food and Culture Supernatants

The selective extraction of the toxin from culture supernatants, serum, stool, and food samples is necessary for detection of the enzymatic activity of BoNT. Clinical samples and foods contain high levels of endogenous proteases that can cleave the peptide substrates or cleave the product peptides before they can be detected by the mass spectrometer (*27*). The selective extraction of very low levels of BoNT from a complex matrix is essential to the Endopep-MS method. Antibodies specific for each BoNT serotype are an efficient way of selectively extracting the toxin from cultures, clinical and food samples (*27*). Antibodies are bound and cross-linked to magnetic beads and added to the culture, serum or a stool extract. The toxin binds to the antibodies that are attached to the magnetic beads. The beads containing antibodies and toxin are removed from the samples and washed to remove materials that were nonspecifically bound (Figure 3). Polyclonal antibodies produced in animals such as rabbits tended to yield higher detection limits for BoNT especially BoNT/A and /B (*27*). We therefore use mixtures of high-affinity monoclonal antibodies from yeast displays that bind selectively to specific epitopes on the BoNT (*28*, *35*). These antibodies were discovered using a yeast library produced from individuals immunized against BoNTs. The panels of antibodies discovered from the yeast displays were then tested in the Endopep-MS method and multiple antibodies for each serotype that yielded the best sensitivity and selectivity were incorporated into the method.

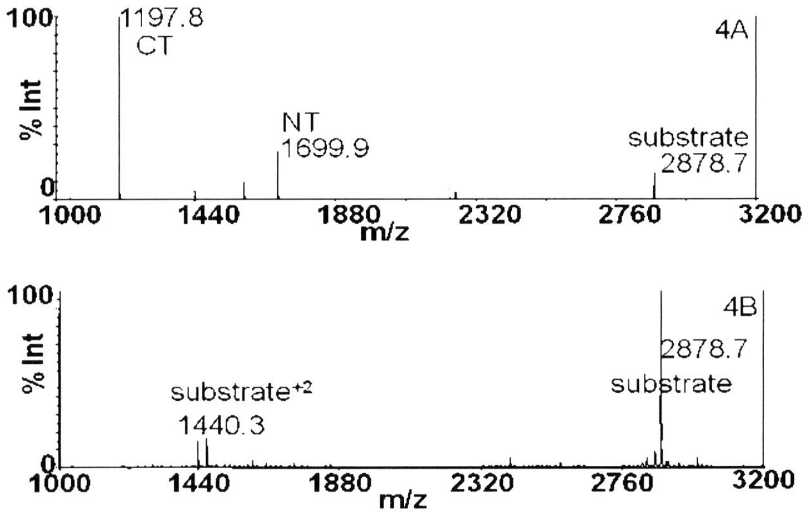

Figure 4. Mass spectrometry detection of the cleavage products for botulinum neurotoxin A (BoNT A). Panel A shows the substrate (Biotin-KGSNRTRIDQGNQRATRLLGGK-Biotin, MH⁺ = 2878.7) is cleaved by BoNT A at a toxin dependent site to form two product peptides. The N-terminal (NT) product peptide is Biotin-KGSNRTRIDQGNQ (MH⁺ = 1699.9 and the C-terminal (CT) product peptide is RATRLLGGK-Biotin (MH⁺ = 1197.8). Panel B is a negative control and shows the singularly and doubly charged substrate ions at m/z 2878.7 and 1140.3 respectively.

Endopep-MS Peptide Cleavage Reactions and Mass Spectrometry Detection

BoNTs are zinc metalloproteases that cleave and inactivate specific cellular proteins essential for the release of the neurotransmitter acetylcholine (Figure 1). Each BoNT serotype recognizes and cleaves a unique site on either SNAP-25 or VAMP-2. We have synthesized peptides that mimic the specific portions of SNAP-25 and VAMP-2 that are substrates for all seven BoNT serotypes (Figure 2) (*26*). The endopeptidase activity is used to detect and differentiate the serotype by allowing the toxin to cleave its specific peptide substrate and detect the cleavage products by mass spectrometry (*25*, *26*). In the presence of active BoNT, the peptide substrate is cleaved in a unique location that is characteristic for each serotype. The reaction mixture is then analyzed by mass spectrometry, which detects the peptides present and accurately reports the mass of each peptide. The presence of the peptide cleavage products corresponding to their toxin-dependent location indicates the presence of a particular serotype of BoNT (*25–30*). If the peptide substrate either remains intact or is cleaved in a location other than the toxin-specific site, that serotype of BoNT is not present above the detection limit of the method. There is no cross-reactivity between the toxin serotypes (*25*, *26*). Each BoNT serotype cleaves only its peptide substrate, and only in the site that is specific for the serotype. Thus, the mass spectral determination of the enzymatic activity differentiates the toxin serotype (*25–30*).

There are advantages to using the specific enzymatic activity to detect and differentiate BoNT. The first major advantage is that the Endopep-MS is an activity-based method that measures enzymatically active toxin only. Protein toxin can be inactivated in several ways, including heat and chemicals, but only active toxins are a threat to human health. Methods such as ELISA tend to measure both active and inactive toxins. The Endopep-MS measures only BoNTs that have their enzymatic activity. The second major advantage is sensitivity. BoNTs are toxic at very low levels because they are efficient enzymes. Measuring the enzymatic activity is an enhancement in the amount of product peptide that is detected. If each BoNT cleaves the substrate peptide 1000 times, then there are three orders of magnitude more product peptide than toxin, yielding much lower limits of detection than could be obtained by detecting the toxin directly. The third major advantage is the specificity of the enzymatic reaction, which allows toxin serotypes to be determined in a single experiment.

Mass Spectrometry Detection of Cleavage Products

The BoNT dependent cleavage reactions are diluted in matrix solution and then deposited on a MALDI plate. After drying, the MALDI plate is placed in the instrument, and spectra of the BoNT dependent cleavage reactions are acquired using a MALDI-TOF mass spectrometer in positive ion, reflector mode. Data is acquired in a mass range of *m/z* 1100 to 4800 to obtain data on the smallest of the cleavage products at *m/z* 1198 up through the largest of the intact substrate peptides at *m/z* 4497. The presence of peaks which correspond to cleavage products indicates the presence of a particular serotype of BoNT. The absence of those peaks indicates the absence of BoNT or the presence of BoNT below the limit of detection. Data acquisition on a MALDI-TOF instrument allows for rapid analysis of each sample, with data acquisition requiring only 15 seconds per sample.

Figure 4A shows the mass spectrum of toxin-dependent cleavage products of the peptide substrate for BoNT/A. Peaks at *m/z* 1198 and 1700 indicate the presence of BoNT/A. These peaks are absent in Figure 4B as this is a negative control. The dominant peak in the negative control is the substrate at *m/z* 2879 as it remains uncleaved.

Figure 5 depicts mass spectra for the reactions of BoNT/B (5A), /E (5B), and /F (5C) with their peptide substrates. Peaks at *m/z* 1760 and 2283 indicate the presence of BoNT/B; 1137 and 2923 are markers for BoNT/E; and 1346 and 3169 indicate BoNT/F.

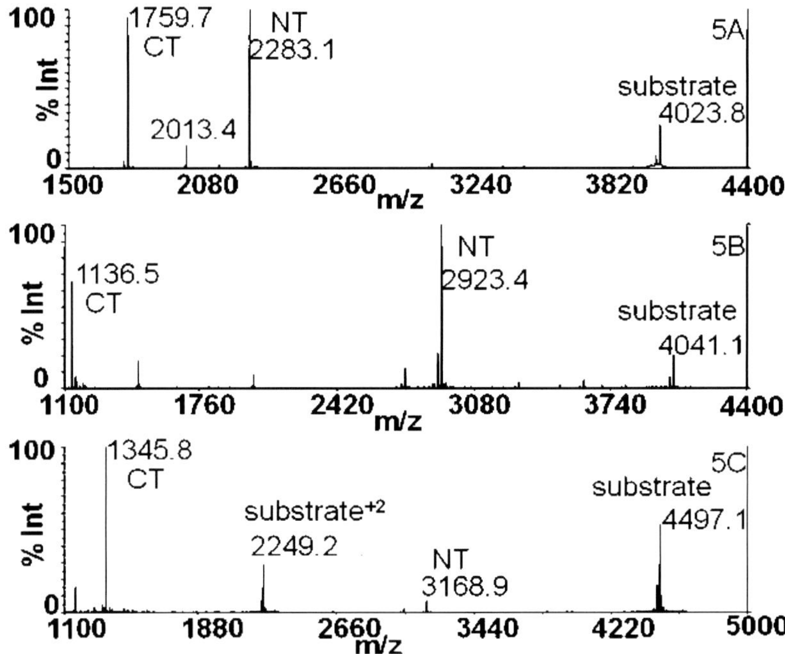

Figure 5. MALDI-TOF mass spectra of the Endopep-MS reaction of BoNT/B (5A), /E (5B), or /F (5C). Peaks at m/z 4024, 4041, and 4497 indicate the presence of the intact peptide substrates for BoNT/B, /E, and /F respectively; cleavage products indicative of BoNT/B are present at m/z 1760 and 2283; 1137 and 2923 are markers for BoNT/E, and peaks at m/z 1346 and 3169 indicate BoNT/F.

Workflow of BoNT Analysis by Endopep-MS

Testing for the presence or absence for BoNT by Endopep-MS is typically accomplished in less than 6 hours as shown in Figure 6, which is a shorter time frame than sample analysis by mouse bioassay which can take 1-4 days. First, samples are aliquoted into 96 well plates, along with all reagents required for the toxin extraction portion of the assay. Next, antibody-coated beads are mixed with the sample for 1 hour. Beads are then washed to reduce non-specific binding, and then the beads are reconstituted in a mixture of the reaction buffer with peptide substrate. This mixture is incubated at 37°C for 4 hours. A portion of the reaction supernatant is then added to matrix solution and spotted on a MALDI plate. As discussed above, the speed of the mass spectrometer allows for rapid analysis of each reaction. Typically, less than 8 samples are analyzed for botulinum neurotoxin at the same time. If the sample size is between 8 and 96 samples, testing time is lengthened by approximately 1 hour.

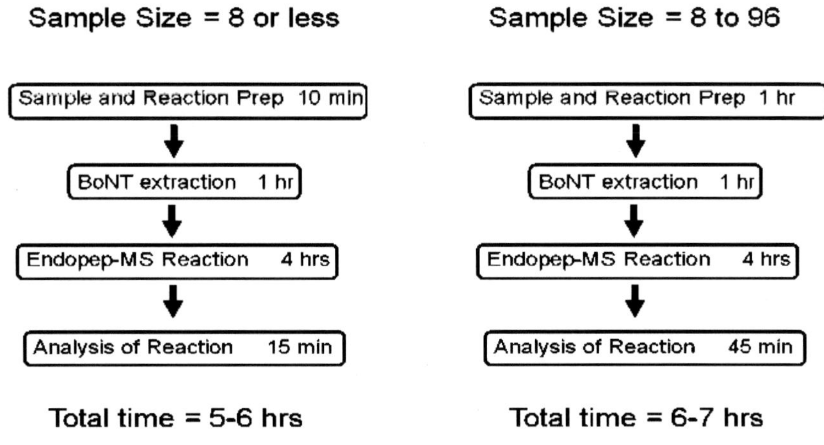

Figure 6. The typical sample workflow for BoNT analysis by Endopep-MS.

Limits of Detection of BoNT in Clinical Samples

As discussed above, BoNT analysis by Endopep-MS is advantageous in terms of speed of the assay. However, speed is not the only advantage of the Endopep-MS method. The limit of detection of toxin with the mouse bioassay is defined as 1 mouse LD_{50} which is thought to be about 10 pg or 67 attomole for BoNT A (*2*). We can achieve limits of detection lower than that of the mouse in most clinical samples. Our work has primarily focused on BoNT/A, /B, /E, and /F as these four serotypes are typically associated with human cases of botulism. Currently, the Endopep-MS method has limits of detection of 0.5, 0.1, 0.1, and 0.1 mouse LD_{50} respectively in human serum. The limits of detection in stool are slightly higher than serum because proteases are more abundant in stool than serum; nonetheless, the current limits of detection of BoNT/A, /B, /E, and /F in stool are 0.5, 0.1, 5, and 0.1 mouse LD_{50} respectively.

Applications of Endopep-MS Assay—Example #1

The limits of detection reported above were obtained with BoNT spiked into matrices. It is important to demonstrate the utility of this method on samples involved in botulism outbreaks in addition to spiked samples. In the summer of 2007, a botulism outbreak was reported in commercially-canned hot dog chili sauce (*36*). We obtained some of the chili sauce extract and tested 0.5 mL of it for BoNT/A, /B, /E, and /F by Endopep-MS. Within 6 hours, we determined that the sample was positive for BoNT/A as evidenced by the presence of peaks at *m/z* 1198 and 1700 in Figure 7C corresponding to cleavage of the peptide substrate by BoNT/A. Figure 7A and 7B are the positive and negative controls. The positive control (Figure 7A) was an extract from a control chili sauce spiked with 1000 mLD_{50} of BoNT A and the negative control is extract of chili sauce with no toxin. The sample was either negative for BoNT/B, /E, and /F or those BoNTs were not present above the limits of detection.

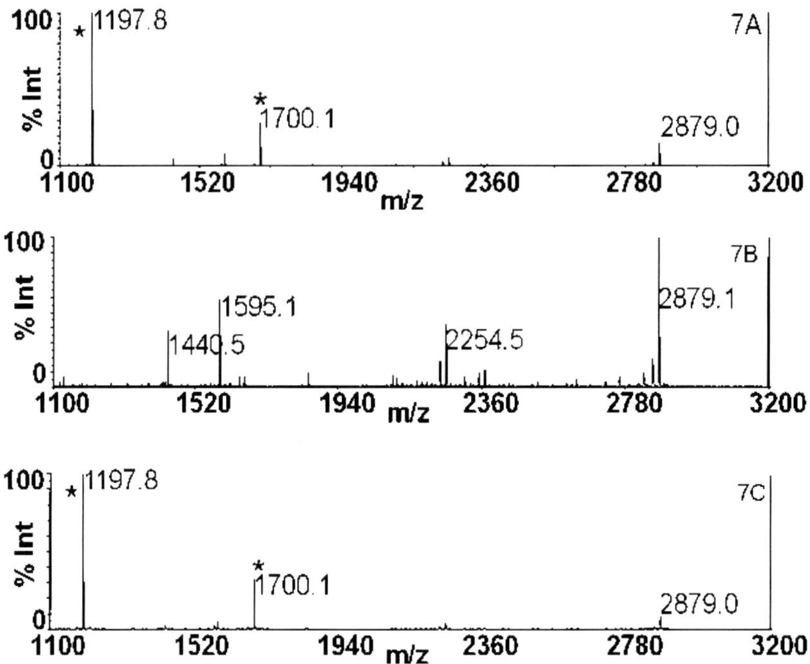

Figure 7. MALDI-TOF mass spectra corresponding to the reaction of chili extract spiked with 1000 mouse LD_{50} of BoNT/A (7A), unspiked chili extract (7B), or chili extract (7C) suspected in a botulism outbreak. Peaks at m/z 1198 and 1700 indicate the presence of BoNT/A in a sample.

Application of Endopep-MS Assay—Example #2

In the spring of 2007, botulism was suspected in catfish in Mississippi (*37*). Serum from the fish was tested by mouse bioassay for the presence of BoNT, and the results were negative. Catfish serum (0.5 mL) was tested for BoNT/A, /B, /E, and /F by Endopep-MS. Within 6 hours the sample was positive for BoNT/E as evidenced by the presence of peaks at *m/z* 1137 and 2923 in Figure 8C corresponding to cleavage of the peptide substrate by BoNT/E. The sample was either negative for BoNT/A, /B, and /F or those BoNTs were present below limits of detection. Although our method as described here does not permit accurate quantification of toxin present in a sample, we estimate that the level of toxin present in the catfish serum to be between 0.01 and 0.5 mouse LD_{50}. This low level of toxin might explain the negative results by mouse bioassay. Figure 8A is the positive control of serum spiked with 0.5 mouse LD_{50} of BoNT/E and Figure 8B is a negative control (blank serum).

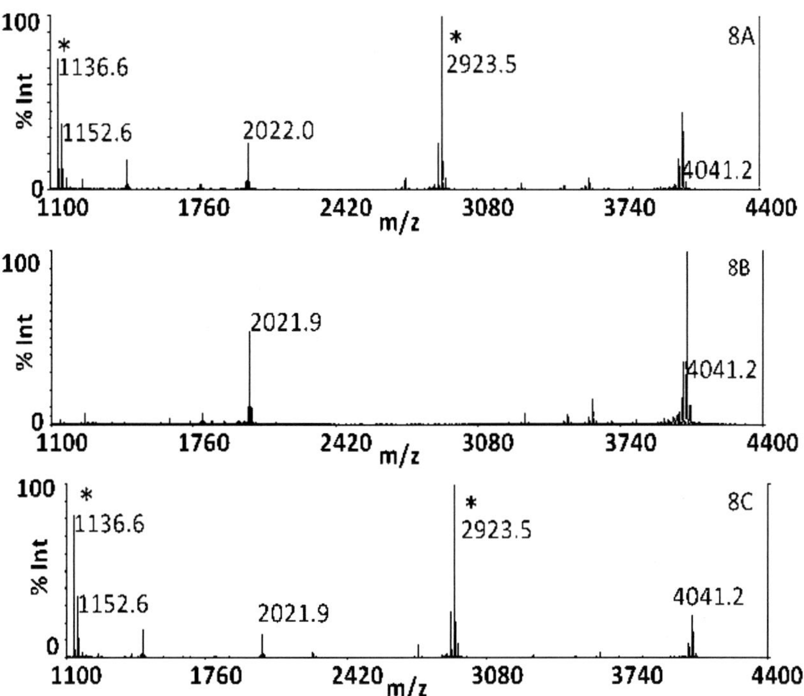

Figure 8. MALDI-TOF mass spectra corresponding to the reaction of serum spiked with 0.5 mouse LD_{50} of BoNT/E (8A), unspiked serum (8B), and catfish serum suspected in a botulism outbreak (8C). Peaks at m/z 1137 and 2923 indicate the presence of BoNT/E in a sample.

Subtyping of BoNT/A

Most of the serotypes of BoNT can further be classified into subtypes, which account for ≥2% amino acid variation between BoNTs of the same serotype. Identification of the subtype can provide useful information for epidemiologists or forensics experts as one can assess the relationship, if any, between concurrent botulism outbreaks. There are currently five recognized subtypes of BoNT/A, known as /A1, /A2, /A3, /A4, and /A5 (*24*, *25*). BoNT/A1 through /A4 have been tested and the toxins have been detected by Endopep-MS (*28*). BoNT/A5 has only recently been discovered and is not yet unavailable for testing. However, the subtypes of BoNT/A cannot be distinguished through Endopep-MS alone as all subtypes of BoNT/A cleave the peptide substrate in the same location, yielding the same data for all subtypes. Therefore, the Endopep-MS method was extended to identify the subtype of BoNT.

```
PFVNKQFNYKDPVNGVDIAYIKIPNVGQMQPVKAFKIHNKIWVIPERDTFTNPEEGDLNP
PPEAKQVPVSYYDSTYLSTDNEKDNYLKGVTKLFERIYSTDLGRMLLTSIVRGIPFWGGS
TIDTELKVIDTNCINVIQPDGSYRSEELNLVIIGPSADIIQFECKSFGHEVLNLTRNGYG
STQYIRFSPDFTFGFEESLEVDTNPLLGAGKFATDPAVTLAHELIHAGHRLYGIAINPNR
VFKVNTNAYYEMSGLEVSFEELRTFGGHDAKFIDSLQENEFRLYYYNKFKDIASTLNKAK
SIVGTTASLQYMKNVFKEKYLLSEDTSGKFSVDKLKFDKLYKMLTEIYTEDNFVKFFKVL
NRKTYLNFDKAVFKINIVPKVNYTIYDGFNLRNTNLAANFNGQNTEINNMNFTKLKNFTG
LFEFYKLLCVRGIITSKTKSLDKGYNKALNDLCIKVNNWDLFFSPSEDNFTNDLNKGEEI
TSDTNIEAAEENISLDLIQQYYLTFNFDNEPENISIENLSSDIIGQLELMPNIERFPNGK
KYELDKYTMFHYLRAQEFEHGKSRIALTNSVNEALLNPSRVYTFFSSDYVKKVNKATEAA
MFLGWVEQLVYDFTDETSEVSTTDKIADITIIIPYIGPALNIGNMLYKDDFVGALIFSGA
VILLEFIPEIAIPVLGTFALVSYIANKVLTVQTIDNALSKRNEKWDEVYKYIVTNWLAKV
NTQIDLIRKKMKEALENQAEATKAIINYQYNQYTEEEKNNINFNIDDLSSKLNESINKAM
ININKFLNQCSVSYLMNSMIPYGVKRLEDFDASLKDALLKYIYDNRGTLIGQVDRLKDKV
NNTLSTDIPFQLSKYVDNQRLLSTFTEYIKNIINTSILNLRYESNHLIDLSRYASKINIG
SKVNFDPIDKNQIQLFNLESSKIEVILKNAIVYNSMYENFSTSFWIRIPKYFNSISLNNE
YTIINCMENNSGWKVSLNYGEIIWTLQDTQEIKQRVVFKYSQMINISDYINRWIFVTITN
NRLNNSKIYINGRLIDQKPISNLGNIHASNNIMFKLDGCRDTHRYIWIKYFNLFDKELNE
KEIKDLYDNQSNSGILKDFWGDYLQYDKPYYMLNLYDPNKYVDVNNVGIRGYMYLKGPRG
SVMTTNIYLNSSLYRGTKFIIKKYASGNKDNIVRNNDRVYINVVVKNKEYRLATNASQAG
VEKILSALEIPDVGNLSQVVVMKSKNDQGITNKCKMNLQDNNGNDIGF
FIGFHQFNNIAKLVASNWYNRQIERSSRTLGCSWEFIPVDDGWGERPL
```

Figure 9. Sequence of botulinum neurotoxin A1 (BoNT/A1 or Hall strain). Underlined residues are mutated in BoNT/A2.

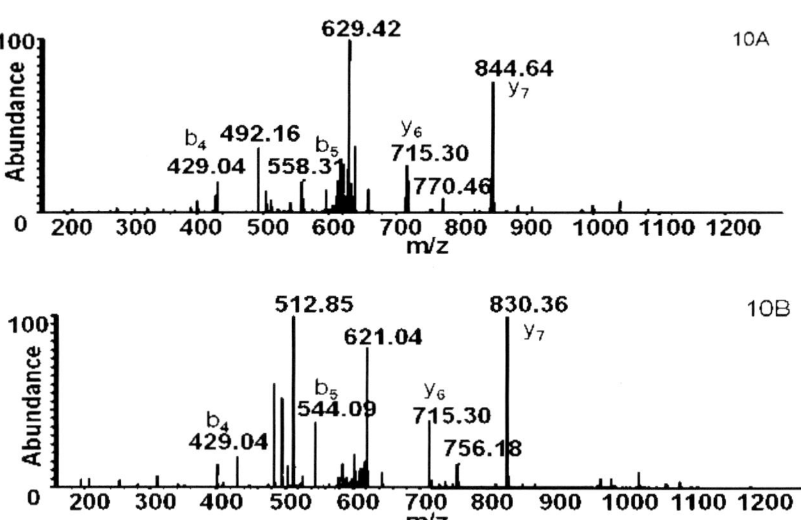

Figure 10. Mass spectra of MS/MS fragmentation of tryptic peptides from BoNT/A1 (10A) and /A2 (10B). The peptide sequence in BoNT/A1 is SFGHEVLNLTR, and the E is mutated to a D in BoNT/A2. This mutation is shown by the altered b_5 and y_7 ions in the spectra.

Following the Endopep-MS reaction, the BoNT attached to the antibody-coated beads is tryptically digested. The resultant tryptic peptides are analyzed by LC-MS/MS. Data are searched against a protein database containing protein sequences of all known subtypes of BoNT. Using this procedure, all the subtypes and strains tested can be differentiated. An example is the differentiation of BoNT/A1 from /A2 (*29*). BoNT/A1 is approximately 90% homologous to /A2; however, one can take advantage of the 10% difference to positively identify BoNT as /A1 or /A2. The amino acid sequence for BoNT A (Hall strain) is shown in Figure 9 with the residues that are different between BoNT/A1 and BoNT/A2 underlined. Through this process, we acquired mass spectrometric evidence for 76 of the known 131 amino acid differences between BoNT/A1 and /A2. An example of one of these 76 amino acid mutations is depicted in Figure 10. Figure 10A represents the MS/MS fragmentation of a tryptic peptide from BoNT/A1 (S^{166}-R^{176}) with the sequence SFGHEVLNLTR. This peptide has a single amino acid mutation in BoNT/A2 where glutamic acid-170 is mutated to an aspartic acid to yield the tryptic peptide from /A2 SFGHDVLNLTR. This amino acid mutation results in a different molecular weight and MS/MS fragmentation pattern, as depicted in Figure 10B. These data indicate that mass spectrometry can be used to distinguish the subtype of BoNT/A. It is important to note that BoNT/A2 is the only known subtype that has the peptide SFGHDVLNLTR.

Conclusions

Diagnosis, treatment and prevention of disease are the central goals of public health. Rapidly identifying botulinum neurotoxins to confirm a diagnosis or discover a source to prevent additional cases of botulism is critical. The Endopep-MS assay is a rapid, sensitive and select method to detect and differentiate BoNT. The method has three levels of selectivity including extraction of the toxin with selective high affinity antibodies, the selective enzymatic activity of the toxin and the use of high resolution MALDI-TOF/MS to specifically identify the cleavage products. The method obtains attomole/mL detection limits because of a toxin dependent amplification of the product peptides due to the BoNT enzymatic activity. Additional information can then be obtained by the proteomic analysis of the tryptic digestion of the extracted toxin. This approach can yield information on the subtype of toxin strain which can help epidemiologists identify similarities or differences in concurrent outbreaks.

References

1. Schiavo, G.; Matteoli, M.; Montecucco, C. *Physiol. Rev.* **2000**, *80* (2), 717–766.
2. Centers for Disease Control and Prevention: Botulism in the United States 1899-1996. *Handbook for Epidemiologists, Clincians, and Laboratory Workers*; Centers for Disease Control and Prevention: Atlanta, GA, 1998.
3. http://www.cdc.gov/nationalsurveillance/PDFs/Botulism_CSTE_2008.pdf.
4. Poulain, B.; Tauc, L.; Maisey, E. A.; Wadsworth, J. D.; Mohan, P. M.; Dolly, J. O. *Proc. Natl. Acad. Sci. U.S.A.* **1988**, *85* (11), 4090–4094.
5. Montal, M. S.; Blewitt, R.; Tomich, J. M.; Montal, M. *FEBS Lett.* **1992**, *313* (1), 12–8.

6. Binz, T. J.; Blasi, S.; Yamasaki, A.; Baumeister, E.; Link, T. C.; Sudhof, R.; Jahn, R.; Niemann, H. *J. Biol. Chem.* **1994**, *269*, 1617–1620.
7. Blasi, J.; Chapman, E. R.; Line, E.; Binz, T.; Yamasaki, S.; De Canilli, P.; Sudhof, T. C.; Niemann, H.; Jahn, R. *Nature* **1993**, *365* (6442), 160–163.
8. Schiavo, G.; Rossetto, O.; Catsicas, S.; Polverino De Laureto, P.; Dasgupta, B. R.; Benfenati, F.; Montecucco, C *J. Biol. Chem.* **1993**, *268*, 23784–23787.
9. Schiavo, G.; Santucci, A.; Dasgupta, B. R.; Mehta, P. P.; Jontes, J.; Benfenati, F.; Wilson, M. C.; Montecucco, C. *FEBS Lett.* **1993**, *335*, 99–103.
10. Foran, P.; Lawrence, G. W.; Shone, C. C.; Foster, K. A.; Dolly, J. O. *Biochemistry* **1996**, *35*, 2630–2636.
11. Williamson, L. C.; Halpern, J. L.; Montecucco, C.; Brown, J. E.; Neale, E. A. *J. Biol. Chem.* **1996**, *271*, 7694–7699.
12. Schiavo, G.; Benfenati, F.; Poulain, B.; Rossetto, O.; Polverino De Laurento, P.; Dasgupta, B. R.; Montecucco, C. *Nature* **1992**, *359*, 832–835.
13. Yamasaki, S.; Baumeister, A.; Binz, T.; Blasi, J.; Link, E.; Cornille, F.; Roques, B.; Fykse, E. M.; Sudhof, T. C.; Jahn, R.; Niemann, H. *J. Biol. Chem.* **1994**, *269*, 12764–12772.
14. Schiavo, G.; Malizio, C.; Trimble, W. S.; Polverino De Laureto, P.; Milan, G.; Sugiyama, H.; Johnson, E. A.; Montecucco, C. *J. Biol. Chem.* **1994**, *269*, 20213–20216.
15. Yamasaki, S.; Binz, T.; Hayashi, T.; Szabo, E.; Yamasaki, N.; Eklund, M.; Jahn, R.; Niemann, H. *Biochem. Biophys. Res. Commun.* **1994**, *200*, 829–835.
16. Blasi, J.; Chapman, E. R.; Yamasaki, S.; Binz, T.; Niemann, H.; Jahn, R. *EMBO J.* **1993**, *12*, 4821–4828.
17. Schiavo, G.; Shone, C. C.; Bennett, M. K.; Scheller, R. H.; Montecucco, C. *J. Biol. Chem.* **1995**, *270*, 10566–10570.
18. Hill, K. K.; Smith, T. J.; Helma, C. H.; Ticknor, L. O.; Foley, B. T.; Svensson, R. T.; Brown, J. L.; Johnson, E. A.; Smith, L. A.; Okinaka, R. T.; Jackson, P. J.; Marks, J. D. *J. Bacteriol.* **2007**, *189* (3), 818–832.
19. Carter, A. T.; Paul, C. J.; Mason, D. R.; Twine, S. M.; Alston, M. J.; Logan, S. M.; Austin, J. W.; Peck, M. W *BMC Genomics* **2009**, *10*, 115–133.
20. Umeda, K.; Seto, Y.; Kohda, T.; Mukamoto, M.; Kozaki, S. *J. Clin. Microbiol.* **2009**, *47* (9), 2720–2728.
21. Wang, X.; Maegawa, T.; Karasawa, T.; Kozaki, S.; Tsukamoto, K.; Gyobu, Y.; Yamakawa, K.; Oguma, K.; Sakaguchi, Y.; Nakamura, S. *Appl. Environ. Microbiol.* **2000**, *66* (11), 4992–7.
22. Chen, Y.; Korkeala, H.; Aarnikunnas, J.; Lindstrom, M. *J. Bacteriol.* **2007**, *189* (23), 8654–50.
23. Kautter, D. A.; Solomon, H. M. *J. Assoc. Anal. Chem.* **1977**, *60*, 541–545.
24. Notermans, S.; Nagel, J. In *Botulinum Neurotoxin and Tetanus Toxin*; Simpson, L. L., Ed.; Academic Press: New York, NY.
25. Barr, J. R.; Moura, H.; Boyer, A. E.; Woolfitt, A. R.; Kalb, S. R.; Pavlopoulos, A.; McWilliams, L. G.; Schmidt, J. G.; Martinez, R. A.; Ashley, D. L. *Emerging Infect. Dis.* **2005**, *11* (10), 1578–1583.

26. Boyer, A. E.; Moura, H.; Woolfitt, A. R.; Kalb, S. R.; Pavlopoulos, A.; McWilliams, L. G.; Schmidt, J. G.; Barr, J. R. *Anal. Chem.* **2005**, *77*, 3916–3924.
27. Kalb, S. R.; Moura, H.; Boyer, A. E.; McWilliams, L. G.; Pirkle, J. L.; Barr, J. R. *Anal. Biochem.* **2006**, *351* (1), 84–92.
28. Kalb, S. R.; Smith, T. J.; Moura, H.; Hill, K.; Lou, J.; Garcia-Rodriguez, C.; Marks, J. D.; Smith, L. A.; Pirkle, J. L.; Barr, J. R. *Int. J. Mass Spectrom.* **2008**, *278*, 101–108.
29. Kalb, S. R.; Goodnough, M. C.; Malizio, C. J.; Pirkle, J. L.; Barr, J. R. *Anal. Chem.* **2005**, *77* (19), 6140–6146.
30. Barr, J. R.; Kalb, S. R.; Moura, H.; Pirkle, J. L. *J. Chem. Health Safety* **2008**, *15* (6), 14–19.
31. Wictome, M.; Newton, K.; Jameson, K.; Hallis, B.; Dunnigan, P.; Mackay, E.; Clarke, S.; Taylor, R.; Gaze, J.; Foster, K.; Shone, C. *Appl. Environ. Microbiol.* **1999**, *65* (9), 3787–3792.
32. Hallis, B.; James, B. A.; Shone, C. C. *J. Clin. Microbiol.* **1996**, *34* (8), 1934–1938.
33. Shone, C. C.; Quinn, C. P.; Wait, R.; Hallis, B.; Fooks, S. G.; Hambleton, P. *Eur. J. Biochem.* **1993**, *7* (3), 965–971.
34. Shone, C. C.; Roberts, A. K. *Eur. J. Biochem.* **1994**, *225* (1), 263–270.
35. Kalb, S. R.; Garcia-Rodriguez, C.; Lou, J.; Baudys, J.; Smith, T. J.; Marks, J. D.; Smith, L. A.; Pirkle, J. L.; Barr, J. R. *PLoS One* **2010**, *5* (8), e12237.
36. Centers for Disease Control and Prevention (CDC). *MMWR Morb Mortal Wkly Rep.* **2007**, *56* (30), 767–769.
37. Gaunt, P. S.; Kalb, S. R.; Barr, J. R. *J. Vet. Diagn. Invest.* **2007**, *19* (4), 349–354.

Chapter 7

Rapid Identification of Food-Borne Pathogens by Top-Down Proteomics Using MALDI-TOF/TOF Mass Spectrometry

Clifton K. Fagerquist*

Produce Safety & Microbiology Research Unit, Western Regional Research Center, Agricultural Research Service, U.S. Department of Agriculture, Albany, California 94710
***clifton.fagerquist@ars.usda.gov**

Rapid identification of bacterial microorganisms is particularly relevant to efforts to monitor the safety and security of domestically grown and imported foods. Mass spectrometry (MS) is increasingly utilized to identify and characterize bacterial microorganisms and in particular food-borne pathogens. Matrix-assisted laser desorption/ionization (MALDI) time-of-flight time-of-flight tandem mass spectrometry (TOF-TOF-MS/MS) has recently been shown to fragment small and modest-sized singly-charged protein ions (without prior protein digestion) to generate sequence-specific fragment ions. These sequence-specific fragment ions can be used for identification of the protein and, if the protein sequence is sufficiently unique, the source microorganism. Our group has developed web-based software for rapid top-down identification of protein biomarkers of bacterial microorganisms from sequence-specific fragment ions analyzed by MALDI-TOF-TOF-MS/MS. The software rapidly compares the mass-to-charge (*m/z*) of MS/MS fragment ions to the *m/z* of *in silico* fragment ions derived from hundreds of bacterial protein sequences that have the same molecular weight as the protein biomarker ion. We have identified several protein biomarkers from pathogenic and non-pathogenic *E. coli* using this top-down proteomic identification approach.

Introduction

Frequent and sporadic outbreaks of food-related illness linked to food-borne pathogens is of increasing concern to public health officials, the food industry and society at large. The frequency and persistence of such outbreaks point to the need to develop rapid, robust and accurate methods to identify and characterize these sometimes deadly microorganisms. Pathogens most frequently associated with outbreaks of food-borne illness are *E. coli* O157:H7 (and other *E. coli* serotypes), *Salmonella, Listeria monocytogenes*, *Campylobacter* and Norovirus. Rapid and unambiguous identification and characterization of these (and other) pathogens is critical for trace-back, clinical and forensic analysis. Because of its speed, sensitivity and specificity, mass spectrometry (MS) is increasingly utilized for rapid microbial identification *via* detection and/or identification of biomolecules that are highly specific for the microorganism which produced them, typically proteins (*1*, *2*) or DNA (*3*). Matrix-assisted laser desorption/ionization time-of-flight mass spectrometry (MALDI-TOF-MS) has become a leading MS platform for microbial identification. Although other MS-based techniques, e.g. electrospray ionization mass spectrometry (ESI-MS) appear to demonstrate greater taxonomic resolution when coupled with liquid chromatography (*4*), MALDI-TOF-MS has advantages of speed, a simpler sample preparation and rapid data analysis (*1*, *2*). Commercial and in-house developed software utilizing pattern recognition algorithms are now routinely used for MALDI-TOF-MS data analysis for the purpose of bacterial identification (*5*). Alternatively, bacterial identification has been demonstrated by correlating the mass-to-charge (*m/z*) ratio of MALDI-TOF-MS peaks to the protein molecular weights (MW) derived from bacterial genomic databases using in-house developed software (*6*).

Tandem TOF instruments (TOF-TOF) were primarily developed for the purpose of high-throughput bottom-up proteomics (*7*, *8*). However, it quickly became apparent that small and modest-sized intact protein ions could also be fragmented, without prior digestion, and that this fragmentation could be used for identification by combining sequence tag and protein MW information to provide a top-down proteomics approach to identification of the protein and its source organism (*9*). It was also noted that the singly or doubly charged (protonated) protein ions generated by MALDI most often fragment due to the number and location of aspartic acid (D), glutamic acid (E) and proline (P) residues (*9*). In the gas phase, acidic acid residues of a protein facilitate fragmentation by transferring a proton from their side-chains to the polypeptide backbone weakening the amide bond. This "top-down" identification strategy was further developed by Demirev and co-workers who demonstrated identification of *Bacillus atrophaeus* and *Bacillus cereus* from MS/MS of intact protein ions (*10*). Using software, developed in-house, the *m/z* of MS/MS fragment ions were compared to a database of *in silico* fragment ions (a-, b-, y- fragment ions and up to two neutral losses: NH_3 and H_2O) derived from bacterial protein sequences. A p-value algorithm, which calculates the probability of a random identification, was developed to score/rank identifications (*10*). This approach does *not* require a sequence "tag" for identification which is important because dissociation of

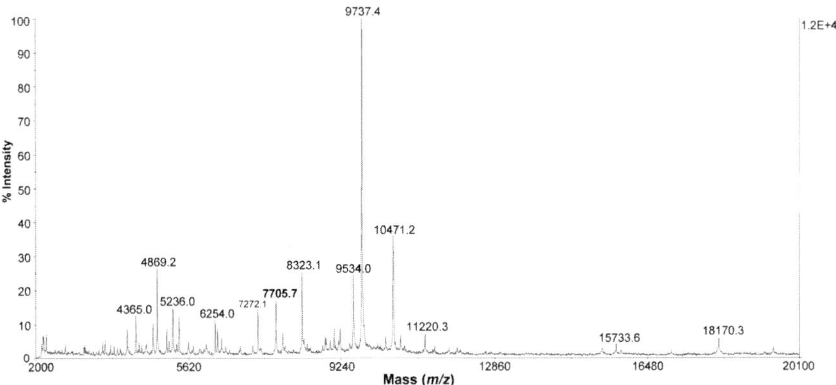

Figure 1. MALDI-TOF-TOF-MS of bacterial cell lysate of E. coli O157:H7 strain RM5603 isolated from water from the Salinas River, California. MALDI matrix: α-cyano-cinnamic acid.

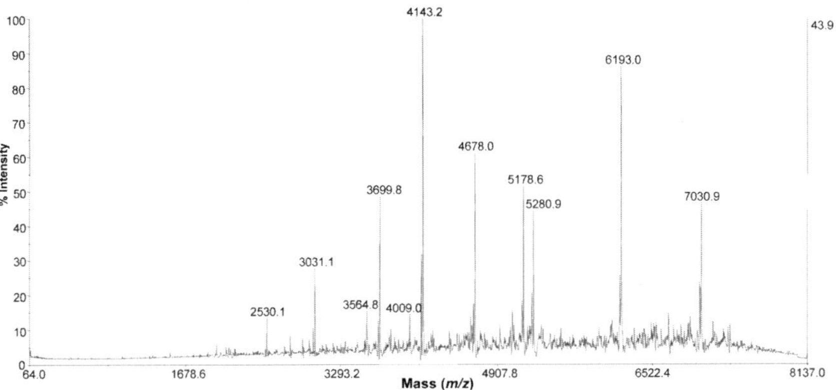

Figure 2. MALDI-TOF-TOF-MS/MS of precursor ion at m/z 7705.7 in Figure 1.

singly charged protein ions do not necessarily generate a series of fragment ions from adjacent residues.

Our laboratory very recently reported using web-based software, developed in-house, to confirm the identity of protein biomarkers of *Campylobacter* species/strains that had previously been identified by bottom-up proteomics techniques (*11*). The software rapidly compares MS/MS fragment ions to *in silico* fragment ions (a, b, b-18, y, y-17, y-18) from hundreds of bacterial proteins sequences. A simple peak-matching algorithm was used to score/rank identifications, and for purposes of comparison Demirev's p-value algorithm was also incorporated into the software to independently score/rank identifications. Both algorithms gave very similar rankings. In addition, our web-based software allows comparison of MS/MS fragment ions to residue-specific fragment ions, e.g. D, E, P residues. We often observed an enhancement of the top identification score (i.e. correct identification) relative to the lower ranking identifications when using a D-, E-, P-specific or D-specific comparison in contrast to a non-residue-specific comparison (*11*).

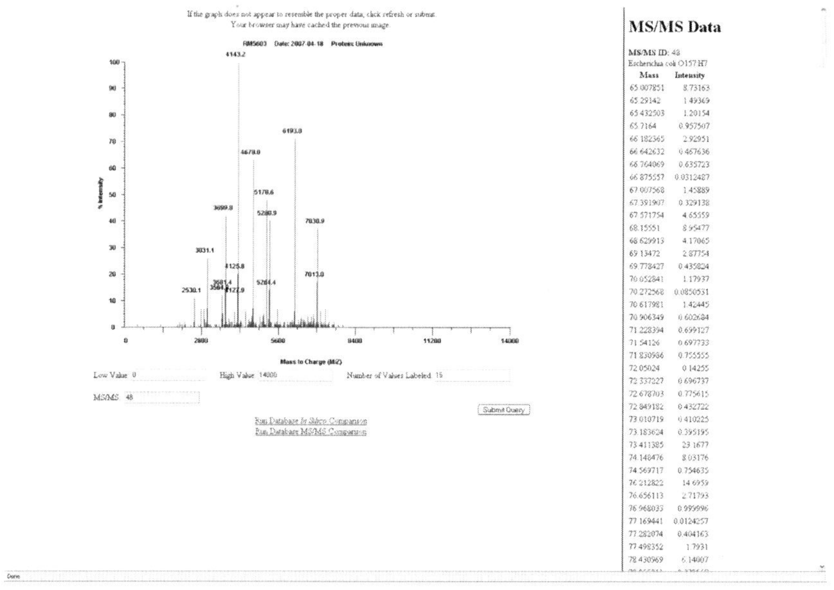

Figure 3. Processed MALDI-TOF-TOF-MS/MS spectra from Figure 2 as displayed in the USDA software.

In the current chapter, results are presented of identification of protein biomarkers from bacterial cell lysates of pathogenic and non-pathogenic *E. coli* using MALDI-TOF-TOF-MS/MS and top-down proteomics. In one case, a single amino acid substitution (D ↔ N) in a protein sequence that results in a protein MW difference of only 1 Da, which is difficult to detect in MS mode, can be easily distinguished by MS/MS and top-down analysis allowing a protein biomarker to be identified unambiguously.

Results and Discussion

Figure 1 shows the MS spectrum of bacterial cell lysate of *E. coli* O157:H7 strain RM5603 analyzed by MALDI-TOF-MS. The peak at *m/z* 7705.7 was analyzed by MALDI-TOF-TOF-MS/MS (Figure 2). The raw MS/MS data was processed and uploaded to the USDA software and compared against *in silico* fragment ions of bacterial protein sequences that have the same MW as that of the protein biomarker ion. Figure 3 is a graphical user interface (GUI) image, as displayed in the USDA software, of the MS/MS data shown in Figure 2. Figure 4 is a GUI image of the "MS/MS-to-*In silico* Comparison" parameter settings window as displayed in the USDA software. MS/MS-to-*in silico* comparison parameters are under operator control, e.g. MS/MS fragment ion minimum intensity threshold (%), the mass range to be compared (*m/z*), protein MW tolerance, fragment ion *m/z* tolerance, a non-residue-specific comparison (blank field) vs. a residue-specific comparison (e.g. D-, E-, P-specific), scoring/ranking algorithm and the number of results to be displayed.

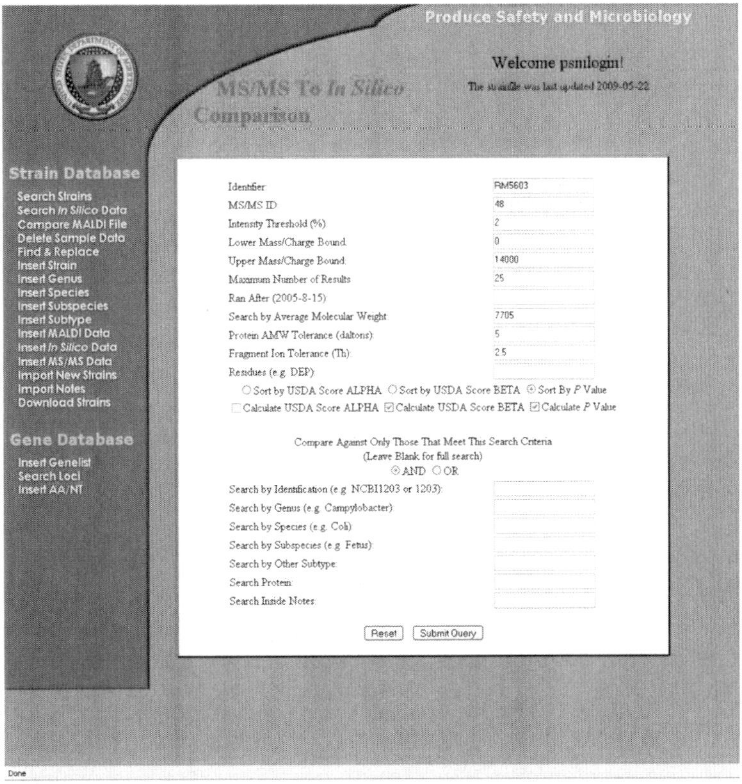

Figure 4. Parameter settings table for MS/MS-to-in silico comparison as displayed in the USDA software.

Table 1A shows the top six identifications of the MS/MS spectrum in Figure 2 using a non-residue-specific comparison. The top identification of both scoring algorithms is the sequence of the YahO protein of *E. coli* O157:H7 strains RM5603 and EDL933 as well as *E. coli* O55:H7 strain RM2057. The 2nd ranked identification is the sequence of the YahO protein of the non-pathogenic/non-O157 *E. coli* strains K-12 and RM3061. The 3rd ranked identifications strongly suggest a random, incorrect identifications. The analysis was performed with a non-residue-specific analysis, i.e. all *in silico* fragment ions (a, b, b-18, y, y-17, y-18) were compared regardless of the amino acid residues adjacent to the site of fragmentation that generated the fragment ion. Table 1B shows the same analysis as that in Table 1A except MS/MS fragment ions were compared to *in silico* fragment ions from cleavage of the polypeptide backbone at sites adjacent to D, E and P residues. The result of this residue-specific analysis is an enhancement of the top identification relative to the runner-up identification (and lower ranked identifications). Note also that the 3rd ranked identifications are different in Tables 1A and 1B which suggests that these identifications are random, incorrect identifications, whereas the 1st and 2nd ranked identifications are the identical in Tables 1A and 1B.

E. coli O157:H7 strains EDL933 and RM5603 and E. coli O55:H7 strain RM2057

MKIIISKMLVGALAFAVTNVYAAELMTKAEFEKVESQYE
KIGDISTSNEMSTADAKEDLIKKADEKGADVLVLTSGQT
DNKIHGTADIYKKK

MW = 9930.41 Da

Mature protein MW = 7707.62 Da

E. coli strain K-12

MKIIISKMLVGALALAVTNVYAAELMTKAEFEKVESQYE
KIGDISTSNEMSTADAKEDLIKKADEKGADVLVLTSGQT
DNKIHGTANIYKKK

MW = 9895.41 Da

Mature protein MW = 7706.64 Da

E. coli strain RM3061

MKIIISKMLVGALAFAVTNVYAAELMTKAEFEKVESQYE
KIGDISTSNEMSTADAKEDLIKKADEKGADVLVLTSGQT
DNKIHGTANIYKKK

MW = 9929.43 Da

Mature protein MW = 7706.64 Da

Figure 5. Amino acid sequence of the putative uncharacterized protein YahO of the pathogenic E. coli O157:H7 strains EDL933 and RM5603, E. coli O55:H7 strain RM2057, the non-O157, non-pathogenic E. coli strains K-12 and RM3061. YahO is post-translationally modified with removal of a 21-residue signal peptide (in outline). Amino acid substitutions between sequences are boxed. The D ↔ N amino acid substitution results in a protein MW difference of 1 Da in the mature protein. Reproduced with permission from reference (12). Copyright 2010.

Figure 5 shows the amino acid sequence of the YahO protein of *E. coli* O157:H7 strains EDL933 and RM5603, *E. coli* O55:H7 strain RM2057, and non-O157 *E. coli* strains K-12 and RM3061. There are two amino acid substitutions between these three sequences. One is located at residue 14 (F ↔ L) in the 21-residue N-terminal signal peptide and, in consequence, is not detected in the mature protein. The other substitution is at residue 65 (D ↔ N) in the mature protein. D (aspartic acid) and N (asparagine) differ in mass by 1 Da. In consequence, this substitution is difficult to detect by MALDI-TOF-MS but is easily detected by MALDI-TOF-TOF-MS/MS. The reason for this is due to the critical rôle that D residues play in fragmentation of singly charged (protonated) protein ions by transferring a proton from their short side-chain to the polypeptide backbone facilitating its fragmentation (*9*). Thus, a single amino acid substitution is easily detectable by MS/MS especially if that substitution involves a D residue. Our laboratory DNA sequenced the *yahO* gene for number of other strains of

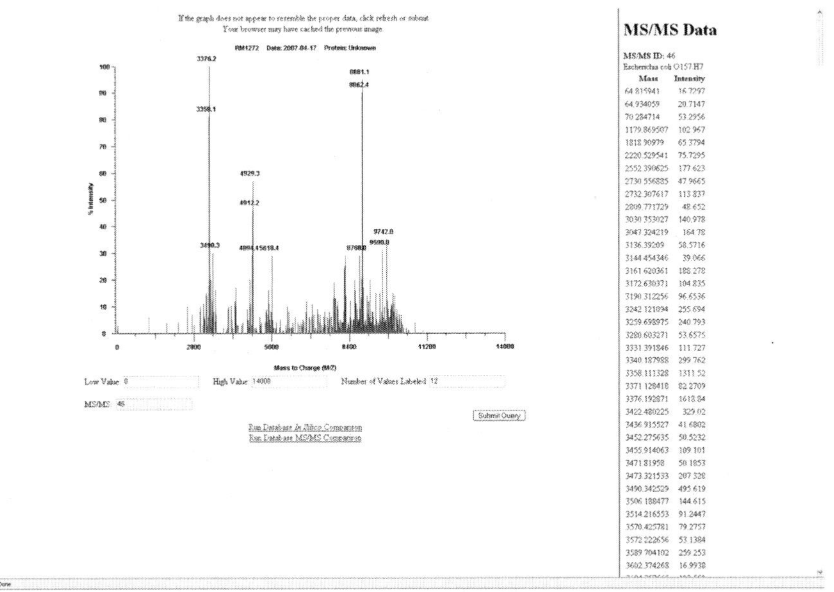

Figure 6. Processed MALDI-TOF-TOF-MS/MS spectra (as displayed in the USDA software) of a protein biomarker ion at m/z ~ 10471.7 from a bacterial cell lysate of E. coli O157:H7 strain EDL933. Low signal-to-noise (S/N) is due to poor fragmentation efficiency of this protein ion.

E. coli including strains of the O55:H7 serotype which is a "near-neighbor" and evolutionary precursor of the more pathogenic O157:H7 serotype. We found that the amino acid sequence of YahO of O55:H7 strains were identical to the YahO sequence of O157:H7 strains (*12*). In consequence, although the YahO sequence can be used to distinguish *E. coli* O157:H7 from other bacterial microorganisms, it cannot, by itself, distinguish between O157:H7 from O55:H7 serotypes (*12*).

As shown in Tables 1A and 1B, algorithm computation times are significantly different. The USDA scoring algorithms are relatively simple formulas for scoring/ranking identifications, whereas Demirev's p-value is a much more mathematically complex scoring algorithm and thus much more computationally intensive. Differences in algorithm computation times are affected by a number of factors including: the number of MS/MS fragment ions, the number of *in silico* protein sequences, non-residue-specific *vs.* residue-specific analysis, etc. When a relatively small number of MS/MS fragment ions are being compared (e.g. 30 or less), computation times of the USDA scores and the p-value are comparable. However, when the number of MS/MS fragment ions increases and/or when a residue-specific analysis is employed, differences in algorithm computation time increases significantly. It should be noted that the p-value is a probability-based calculation and a p-value score indicates the probability that an identification occurred randomly (*10*). In contrast, the USDA-β scoring algorithm calculates only the percentage matched MS/MS fragment ions (*11*). Interestingly, the USDA and the p-value algorithms, for the most part, give quite similar rankings.

E. coli O157:H7 strain EDL933 and E. coli O55:H7 strain RM2057

MKKMTKLATLFLTATLSLASGAALAADSGAQTNNGQAN
AAADAGQVAPDARENVAPNNVDNNGVNTGSGGTMLHP
DGSSMNNDGMTKDEEHKNTMCKDGRCPDINKKVQTGD
S············S
GINNDVDTKTDGTTQ

MW = 12882.06 Da
Mature protein MW = 10473.06 Da

E. coli strain K-12, E. coli O55 strain RM7208 and S. flexneri

MKKMTKLATLFLTATLSLASGAALAADSGAQTNNGQAN
AAADAGQVAPDARENVAPNNVDNNGVNTGSGGTMLHS
DGSSMNNDGMTKDEEHKNTMCKDGRCPDINKKVQTGD
S············S
GINNDVDTKTDGTTQ

MW = 12872.02 Da
Mature protein MW = 10463.02 Da

Figure 7. Amino acid sequence of the putative homeobox (or YbgS) protein of E. coli O157:H7 strain EDL933, E. coli O55:H7 strain RM2057, E. coli strain K-12, E. coli O55 strain RM7208 and S. flexneri. Sequences are post-translationally modified with a 24-residue N-terminal signal peptide (in outline) and a disulfide bond (S····S) between two cysteine residues. Amino acid substitutions between sequences are boxed. Reproduced with permission from reference (12). Copyright 2010.

MKKVLGVILGGLLLLPVVSNAADAQKAADNKKPVNSW
TCEDFLAVDESFQPTAVGFAEALNNKDKPEDAVLDVQGI
S·····················S
ATVTPAIVQACTQDKQANFKDKVKGEWDKIKKDM

MW = 11857.65 Da
Mature protein MW = 9738.91 Da

Figure 8. Amino acid sequence of the acid stress chaperone-like protein HdeA whose sequence is conserved between E. coli O157:H7 strains EDL933 and RM5603, E. coli strain O55:H7 strain RM2057, E. coli strain K-12 and RM3061 and S. flexneri. The mature protein is post-translationally modified with removal of a 21-residue signal peptide (in outline) and a putative disulfide bond (S····S) between the two cysteine residues.

The peak at *m/z* 10471.2 observed in Figure 1 was also observed in the MS spectrum of the bacterial cell lysate of *E. coli* O157:H7 strain EDL933 at *m/z* 10471.7. This biomarker was analyzed by MS/MS and its spectrum is displayed in Figure 6 (as viewed in the USDA software). The MS/MS spectrum is quite noisy with poor S/N. In consequence, a higher minimum intensity threshold cutoff was used (8%) for comparing MS/MS fragment ions to *in silico* fragment ions. As shown in Table 2, the top ranked identification of both algorithms using a non-residue-specific comparison is the putative homeobox protein sequence of *E. coli* O157:H7 strains EDL933 and RM5603 and *E. coli* O55:H7 strain RM2057. A homeobox protein is a DNA-binding protein involved in gene regulation. The 2nd ranked identifications of the USDA algorithm are the YbgS protein of *E. coli* O55 strain RM7208 and *E. coli* strains K-12 or *S. flexneri*. The poor S/N of the MS/MS spectrum resulted in top identification scores that are relatively close to that of the lower ranked identifications, however the fact that a correct identification was still obtained analyzing poor quality MS/MS data suggests the robustness of this approach for top-down identification.

Figure 7 shows the amino acid sequence of the homeobox protein of *E. coli* O157:H7 strains EDL933 and RM5603, and the YbgS of *E. coli* strain K-12, *E. coli* O55 strain RM7208 and *S. flexneri*. The homebox/YbgS protein has a 24-residue N-terminal signal peptide and a disulfide bond. A single amino acid variation at residue 50 (P ↔ S) in the mature protein results in a protein MW difference of 10 Da. This variation also allows these two sequences to be distinguished from one another by MS/MS and top-down proteomics. Once again, subsequent DNA sequencing revealed that the amino acid sequence for homeobox/YbgS protein from strains of O55:H7 serotype were identical to those of the O157:H7 serotype (*12*).

The peak at *m/z* 9737.4 shown in Figure 1 was also observed in the MS spectrum of the bacterial cell lysate of *E. coli* O157:H7 strain EDL933 at *m/z* 9737.5. This biomarker was analyzed by MS/MS and top-down proteomics and identified as the acid stress chaperone-like protein HdeA sequence of *E. coli* O157:H7 strains EDL933 and RM5603 and *E. coli* O55:H7 strain RM2057 (Table 3). However, the sequence of this protein is identical to the sequence of HdeA of *E. coli* strains K-12 and RM3061 and *S. flexneri*, and thus is not a "good" biomarker to distinguish pathogenic from non-pathogenic strains of *E. coli*. In addition, the O157:H7 HdeA sequence is identical to the HdeA sequence for the O55:H7 serotype (*12*). HdeA is post-translationally modified with a 21-residue N-terminal signal peptide and a disulfide bond (Figure 8).

Previous research, by DNA sequencing and bottom-up proteomics, identified a peak at *m/z* ~ 9060, observed in the MS spectra of non-pathogenic, non-O157:H7 strains (but absent from MS spectra *E. coli* O157:H7 strains), as the acid stress chaperone-like protein HdeB (*13–15*). A non-pathogenic, non-O157:H7 *E. coli* strain RM3061, isolated from Romaine lettuce as part of a USDA environmental survey, was analyzed by MALDI-TOF-TOF-MS. A peak detected at *m/z* 9063.4 in the MS spectrum was analyzed by MS/MS and top-down proteomics and identified as HdeB of non-pathogenic, non-O157:H7 *E. coli* strains RM3061 and K-12 (Table 4). However, the sequence is also shared with strains of the *E. coli* serotypes O55:H7, O55:H6, O55:HN, O6 and *S. flexneri*. HdeB is also post-translationally modified with removal of a N-terminal signal peptide and a disulfide bond (Figure 9).

MNIISSLRKAFIFMGAVAALSLVNAQSALA**ANESAKDMTC**

·······S

QEFIDLNPKAMTPVAWWMLHEETVYKGGDTVTLNETD

S·······

LTQIPKVIEYCKKNPQKNLYTFKNQASNDLPN

MW = 12042.86

Mature protein MW = 9063.26

Figure 9. Amino acid sequence of the acid-stress chaperone-like protein HdeB whose sequence is conserved between E. coli strains RM3061, K-12, E. coli O55:H7 strain RM2057, E. coli O55:H6 strain RM2068, E. coli O55:HN strain RM2024, E. coli O6 and S. flexneri. The mature protein is post-translationally modified with removal of a 29-residue signal peptide (in outline) and a putative disulfide bond (····S S·····) between the two cysteine residues.

Conclusions

Results have been presented which demonstrate identification of protein biomarkers of pathogenic and non-pathogenic *E. coli* from bacterial cell lysates of pure cultures by fragmentation of singly charged protein ions using MALDI-TOF-TOF-MS/MS and top-down proteomics. Protein biomarker identification may also provide unambiguous identification of the microorganism if the protein sequence is sufficiently unique. A single amino acid substitution between the YahO sequence of *E. coli* O157:H7 strains RM5603 and EDL933 *vs. E. coli* strain K-12 (a non-pathogenic, non-O157 *E. coli*) is sufficient by MALDI-TOF-TOF-MS/MS and top-down proteomics to distinguish between these two microorganisms. However, the YahO sequence of *E. coli* O157:H7 is also fully homologous to the YahO sequence of *E. coli* O55:H7, a "near-neighbor" serotype of O157:H7. The ability to unambiguously identify a microorganism from a single protein sequence depends on its uniqueness. This issue is particularly relevant when attempting to distinguish between closely-related microorganisms where there may be significant sequence homology among high copy proteins. Protein sequence homology is, of course, affected by the extent of genetic diversity within (and across) genera, species, sub-species, serotype/serovar and strains of a microorganism. Microorganisms that are genetically promiscuous (e.g. *Campylobacter*) are more likely to be unambiguously identified from the sequence of a single protein biomarker than a microorganism whose genome is relatively homogenous and invariant over time.

The software utilized in this work is available free of charge to other researchers with execution of an appropriate control usage agreement.

Table 1A. Top identification scores of a protein biomarker ion at *m/z* 7705.7 (Figures 1 & 2) analyzed by top-down proteomics using a non-residue-specific *in silico* fragment ion comparison. Reproduced with permission from reference (*12*). Copyright 2010

In Silico ID	*Identifier*	*Sample Name*	*Protein*	*USDA Score*	*P-value*
43989	**>◊|WGM|WGM_PSMRU_5A**	***Escherichia coli* O157:H7 (strain RM5603)**	**YahO protein PTM-21SigPep 7707.62**	**61.67**	**6.7E-19**
26947	>tr|Q8X699|Q8X699_ECO57	*Escherichia coli* O157:H7 (strain EDL933)	Putative uncharacterized protein YahO PTM-21SigPep 7707.62	61.67	6.7E-19
43962	>◊|WGM|WGM_PSMRU_1A	*Escherichia coli* O55:H7 (strain RM2057)	YahO protein PTM-21SigPep 7707.62	61.67	6.7E-19
26281	>sp|P75694|YAHO_ECOLI	*Escherichia coli* (strain K-12)	UPF0379 protein YahO PTM-21SigPep 7706.64	57.50	1.7E-15
43983	>◊|WGM|WGM_PSMRU_2A	Non-O157:H7 *Escherichia coli* (strain RM3061)	YahO protein PTM-21SigPep 7706.64	57.50	1.7E-15

Continued on next page.

Table 1A. (Continued). Top identification scores of a protein biomarker ion at *m/z* 7705.7 (Figures 1 & 2) analyzed by top-down proteomics using a non-residue-specific *in silico* fragment ion comparison.

In Silico ID	*Identifier*	*Sample Name*	*Protein*	*USDA Score*	*P-value*
25669	>tr|A5V415|A5V415_SPHWW	*Sphingomonas wittichii* (strain RW1 / DSM 6014 / JCM 10273)	DNA binding domain, excisionase family 7703.11	35.00	
25710	>tr|A6Y3Y9|A6Y3Y9_VIBCH	*Vibrio cholerae* RC385	Transcriptional regulator 7704.84		3.8E-3

MS/MS to *in silico* comparison parameters
Intensity threshold: 2%
Number of MS/MS peaks with intensity ≥ 2%: 120.
m/z range for comparison: 0-14,000 Th.
Fragment ion tolerance: 2.5 Th.
Protein MW 7705 ± 10 Da. Number of bacterial proteins 1323.
All *in silico* fragment ions compared.
"PTM N-Met" indicates that the *in silico* protein sequence was modified to remove the N-terminal methionine.
"PTM #SigPep" indicates that the *in silico* protein sequence was modified to remove a signal peptide.

Algorithm computation times
USDA peak matching algorithm: 43.0 seconds.
P-value: 189.6 seconds.

Table 1B. Top identification scores of a protein biomarker ion at *m/z* 7705.7 (Figures 1 & 2) analyzed by top-down proteomics using a D-, E-, P-specific *in silico* fragment ion comparison. Reproduced with permission from reference (*12*). Copyright 2010

In Silico ID	*Identifier*	*Sample Name*	*Protein*	*USDA Score*	*P-value*
43989	**>◊|WGM|WGM_PSMRU_5A**	***Escherichia coli* O157:H7 (strain RM5603)**	**YahO protein PTM-21SigPep 7707.62**	**41.67**	**5.3E-19**
26947	>tr|Q8X699|Q8X699_ECO57	*Escherichia coli* O157:H7 (strain EDL933)	Putative uncharacterized protein YahO PTM-21SigPep 7707.62	41.67	5.3E-19
43962	>◊|WGM|WGM_PSMRU_1A	*Escherichia coli* O55:H7 (strain RM2057)	YahO protein PTM-21SigPep 7707.62	41.67	5.3E-19
26281	>sp|P75694|YAHO_ECOLI	*Escherichia coli* (strain K-12)	UPF0379 protein YahO PTM-21SigPep 7706.64	34.17	2.0E-13

Continued on next page.

Table 1B. (Continued). Top identification scores of a protein biomarker ion at *m/z* 7705.7 (Figures 1 & 2) analyzed by top-down proteomics using a D-, E-, P-specific *in silico* fragment ion comparison.

In Silico ID	*Identifier*	*Sample Name*	*Protein*	*USDA Score*	*P-value*
43983	>◊|WGM|WGM_PSMRU_2A	Non-O157:H7 *Escherichia coli* (strain RM3061)	YahO protein PTM-21SigPep 7706.64	34.17	2.0E-13
25880	>tr|B1SUG8|B1SUG8_9BACI	*Geobacillus* WCH70	Putative uncharacterized protein 7707.91	22.50	1.4E-4

MS/MS to *in silico* comparison parameters
Intensity threshold: 2%
Number of MS/MS peaks with intensity ≥ 2%: 120.
m/z range for comparison: 0-14,000 Th.
Fragment ion tolerance: 2.5 Th.
Protein MW 7705 ± 10 Da. Number of bacterial proteins 1323.
D-, E-, P-specific *in silico* fragment ions compared.
"PTM N-Met" indicates that the *in silico* protein sequence was modified to remove the N-terminal methionine.
"PTM #SigPep" indicates that the *in silico* protein sequence was modified to remove a signal peptide.

Algorithm computation times
USDA peak matching algorithm: 15.6 seconds.
P-value: 274.0 seconds.

Table 2. Top identification scores of a protein biomarker ion at *m/z* ~10472 observed in the MS spectrum of *E. coli* O157:H7 strain EDL933 analyzed by MS/MS and top-down proteomics using a non-residue-specific *in silico* fragment ion comparison. The relative value of the top identification score is adversely affected by the low S/N of the MS/MS spectrum analyzed (Figure 6). Reproduced with permission from reference (*12*). Copyright 2010

In Silico ID	*Identifier*	*Sample Name*	*Protein*	*USDA Score*	*P-value*
36101	**>tr|Q8X948|Q8X948_ECO57**	***Escherichia coli* O157:H7 (strain EDL933)**	**Putative homeobox protein PTM-24SigPep 10473.06**	**41.58**	**2.2E-4**
43990	>◊|WGM|WGM_PSMRU_5B	*Escherichia coli* O157:H7 (strain RM5603)	YbgS (homeobox) protein PTM-24SigPep 10473.06	41.58	2.2E-4
43973	>◊|WGM|WGM_PSMRU_1B	*Escherichia coli* O55:H7 (strain RM2057)	YbgS (or homeobox) protein PTM-24SigPep 10473.06	41.58	2.2E-4
43995	>◊|WGM|WGM_PSMRU_7B	*Escherichia coli* O55 (strain RM7208)	YbgS (or homeobox) protein PTM-24SigPep 10463.02	38.61	
32552	>tr|A6FIF8|A6FIF8_9GAMM	*Moritella* PE36	Putative uncharacterized protein PTM Met 10472.03		1.1E-3
35345	>sp|P0AAV6|YBGS_ECOLI	*Escherichia coli* (strain K-12)	Uncharacterized YbgS protein PTM-24SigPep 10463.02	38.61	1.9E-3

Continued on next page.

Table 2. (Continued). Top identification scores of a protein biomarker ion at *m/z* ~10472 observed in the MS spectrum of *E. coli* O157:H7 strain EDL933 analyzed by MS/MS and top-down proteomics using a non-residue-specific *in silico* fragment ion comparison. The relative value of the top identification score is adversely affected by the low S/N of the MS/MS spectrum analyzed (Figure 6).

In Silico ID	*Identifier*	*Sample Name*	*Protein*	*USDA Score*	*P-value*
35346	>sp|P0AAV7|YBGS_SHIFL	*Shigella flexneri*	Uncharacterized YbgS protein PTM-24SigPep 10463.02	38.61	
43995	>◊|WGM|WGM_PSMRU_7B	*Escherichia coli* O55 (strain RM7208)	YbgS (or homeobox) protein PTM-24SigPep 10463.02		1.9E-3
32552	>tr|A6FIF8|A6FIF8_9GAMM	*Moritella* PE36	Putative uncharacterized protein PTM Met 10472.03	37.62	
35346	>sp|P0AAV7|YBGS_SHIFL	*Shigella flexneri*	Uncharacterized YbgS protein PTM-24SigPep 10463.02		1.9E-3

MS/MS to *in silico* comparison parameters
Intensity threshold: 8%
Number of MS/MS peaks with intensity ≥ 8%: 101.
m/z range for comparison: 0-14,000 Th.
Fragment ion tolerance: 2.5 Th.

Protein MW 10471 ± 10 Da. Number of bacterial proteins 2041.
All *in silico* fragment ions compared.
"PTM N-Met" indicates that the *in silico* protein sequence was modified to remove the N-terminal methionine.
"PTM #SigPep" indicates that the *in silico* protein sequence was modified to remove a signal peptide.

Algorithm computation times
USDA peak matching algorithm: 71.4 seconds.
P-value: 204.8 seconds.

Table 3. (ID #25) Top identifications scores of a protein biomarker ion at *m/z* 9737.5 observed in the MS spectrum of *E. coli* O157:H7 strain RM1272 (EDL933) analyzed by MS/MS and top-down proteomics using a non-residue-specific *in silico* fragment ion comparison. Reproduced with permission from reference (*12*). Copyright 2010

<table>
<tr><th>In Silico ID</th><th>Identifier</th><th>Sample Name</th><th>Protein</th><th>USDA Score</th><th>P-value</th></tr>
<tr><td>24512</td><td>>sp|P0AET0|HDEA_ECO57</td><td>Escherichia coli O157:H7 (strain EDL933)</td><td>Chaperone-like protein HdeA PTM-21SigPep 9738.91</td><td>52.50</td><td>2.7E-8</td></tr>
<tr><td>43991</td><td>>◊|WGM|WGM_PSMRU_5C</td><td>Escherichia coli O157:H7 (strain RM5603)</td><td>HdeA acid stress chaperone-like protein PTM_21SigPep 9738.91</td><td>52.50</td><td>2.7E-8</td></tr>
<tr><td>43979</td><td>>◊|WGM|WGM_PSMRU_1C</td><td>Escherichia coli O55:H7 (strain RM2057)</td><td>HdeA acid stress chaperone protein PTM-21SigPep 9738.91</td><td>52.50</td><td>2.7E-8</td></tr>
<tr><td>24511</td><td>>sp|P0AES9|HDEA_ECOLI</td><td>Escherichia coli (strain K-12)</td><td>Chaperone-like protein HdeA PTM-21SigPep 9738.91</td><td>52.50</td><td>2.7E-8</td></tr>
<tr><td>43985</td><td>>◊|WGM|WGM_PSMRU_2C</td><td>Non-O157:H7 Escherichia coli (strain RM3061)</td><td>HdeA acid resitance chaperone protein PTM-21SigPep 9738.91</td><td>52.50</td><td>2.7E-8</td></tr>
<tr><td>24513</td><td>>sp|P0AET1|HDEA_SHIFL</td><td>Shigella flexneri</td><td>Chaperone-like protein HdeA PTM-21SigPep 9738.91</td><td>52.50</td><td>2.7E-8</td></tr>
</table>

<table>
<tr><th>In Silico ID</th><th>Identifier</th><th>Sample Name</th><th>Protein</th><th>USDA Score</th><th>P-value</th></tr>
<tr><td>25275</td><td>>tr|B3E2H3|B3E2H3_GEOLS</td><td>Geobacter lovleyi (strain ATCC BAA-1151 / DSM 17278 / SZ)</td><td>Anti-sigma-28 factor, FlgM PTM N-Met 9734.00</td><td>40.00</td><td></td></tr>
<tr><td>24594</td><td>>tr|Q7VNM3|Q7VNM3_HAEDU</td><td>Haemophilus ducreyi</td><td>Putative uncharacterized protein PTM-N-Met 9743.18</td><td></td><td>1.2E-3</td></tr>
</table>

MS/MS to *in silico* comparison parameters
Intensity threshold: 2%.
Number of MS/MS peaks with intensity $\geq$ 2%: 80.
m/z range for comparison: 0-14,000 Th.
Fragment ion tolerance: 2.5 Th.
Protein MW 9737 $\pm$ 10 Da. Number of bacterial proteins 2017.
All *in silico* fragment ions compared.
"PTM N-Met" indicates that the *in silico* protein sequence was modified to remove the N-terminal methionine.
"PTM #SigPep" indicates that the *in silico* protein sequence was modified to remove a signal peptide.

Algorithm computation times
USDA peak matching algorithm: 59.7 seconds.
P-value: 123.6 seconds.

Table 4. (ID #50) Top identification scores of a protein biomarker ion at *m/z* 9063.4 observed in the MS spectrum of a non-pathogenic, non-O157:H7 *E. coli* strain RM3061 and analyzed by MS/MS and top-down proteomics using a non-residue-specific *in silico* fragment ion comparison. Reproduced with permission from reference (*12*). Copyright 2010

In Silico ID	*Identifier*	*Sample Name*	*Protein*	*USDA Score*	*P-value*
43963	**>◊|WGM|WGM_PSMRU_2D**	**Non-O157:H7 *Escherichia coli* (strain RM3061)**	**HdeB acid stress chaperone protein PTM-29SigPep 9063.26**	**52.17**	**8.2E-10**
33855	>sp|P0AET2|HDEB_ECOLI	*Escherichia coli* (strain K-12)	Protein HdeB PTM-29SigPep 9063.26	52.17	8.2E-10
43980	>◊|WGM|WGM_PSMRU_1D	*Escherichia coli* O55:H7 (strain RM2057)	HdeB acid stress chaperone protein PTM-29SigPep 9063.26	52.17	8.2E-10
43969	>◊|WGM|WGM_PSMRU_3D	*Escherichia coli* O55:H6 (strain RM2068)	HdeB acid stress chaperone protein PTM-29SigPep 9063.26	52.17	8.2E-10
43976	>◊|WGM|WGM_PSMRU_4D	*Escherichia coli* O55:HN (strain RM2024)	HdeB acid stress chaperone protein PTM-29SigPep 9063.26	52.17	8.2E-10
33856	>sp|P0AET3|HDEB_ECOL6	*Escherichia coli* O6	Protein HdeB PTM-29SigPep 9063.26	52.17	8.2E-10

<table>
<tr><th>In Silico ID</th><th>Identifier</th><th>Sample Name</th><th>Protein</th><th>USDA Score</th><th>P-value</th></tr>
<tr><td>33857</td><td>>sp|P0AET4|HDEB_SHIFL</td><td>Shigella flexneri</td><td>Protein HdeB
PTM-29SigPep
9063.26</td><td>52.17</td><td>8.2E-10</td></tr>
<tr><td>34374</td><td>>tr|Q15UH4|Q15UH4_PSEA6</td><td>Pseudoalteromonas atlantica (strain T6c / BAA_1087)</td><td>Putative uncharacterized
protein PTM-Met
9062.05</td><td>38.04</td><td>1.6E-3</td></tr>
</table>

MS/MS to *in silico* comparison parameters
Intensity threshold: 2%.
Number of MS/MS peaks with intensity $\geq$ 2%: 92.
m/z range for comparison: 0-14,000 Th.
Fragment ion tolerance: 2.5 Th.
Protein MW 9063 $\pm$ 10 Da. Number of bacterial proteins 1450.
All *in silico* fragment ions compared.
"PTM N-Met" indicates that the *in silico* protein sequence was modified to remove the N-terminal methionine.
"PTM #SigPep" indicates that the *in silico* protein sequence was modified to remove a signal peptide.

Algorithm computation times
USDA peak matching algorithm: 42.1 seconds.
P-value: 115.6 seconds.

References

1. Fenselau, C.; Demirev, P. A. *Mass Spectrom. Rev.* **2001**, *20*, 157–171.
2. Lay, J. O., Jr. *Mass Spectrom. Rev.* **2001**, *20*, 172–194.
3. Ecker, D. J.; Sampath, R.; Massire, C.; Blyn, L. B.; Hall, T. A.; Eshoo, M. W.; Hofstadler, S. A. *Nat. Rev. Microbiol.* **2008**, *6*, 553–558.
4. Williams, T. L.; Monday, S. R.; Feng, P. C.; Musser, S. M. *J. Biomol. Tech.* **2005**, *16*, 134–142.
5. Wunschel, S. C.; Jarman, K. H.; Petersen, C. E.; Valentine, N. B.; Wahl, K. L.; Schauki, D.; Jackman, J.; Nelson, C. P.; White, E. t. *J. Am. Soc. Mass Spectrom.* **2005**, *16*, 456–462.
6. Demirev, P. A.; Lin, J. S.; Pineda, F. J.; Fenselau, C. *Anal. Chem.* **2001**, *73*, 4566–4573.
7. Medzihradszky, K. F.; Campbell, J. M.; Baldwin, M. A.; Falick, A. M.; Juhasz, P.; Vestal, M. L.; Burlingame, A. L. *Anal. Chem.* **2000**, *72*, 552–558.
8. Suckau, D.; Resemann, A.; Schuerenberg, M.; Hufnagel, P.; Franzen, J.; Holle, A. *Anal. Bioanal. Chem.* **2003**, *376*, 952–965.
9. Lin, M.; Campbell, J. M.; Mueller, D. R.; Wirth, U. *Rapid Commun. Mass Spectrom.* **2003**, *17*, 1809–1814.
10. Demirev, P. A.; Feldman, A. B.; Kowalski, P.; Lin, J. S. *Anal. Chem.* **2005**, *77*, 7455–7461.
11. Fagerquist, C. K.; Garbus, B. R.; Williams, K. E.; Bates, A. H.; Boyle, S.; Harden, L. A. *Appl. Environ. Microbiol.* **2009**, *75*, 4341–4353.
12. Fagerquist, C. K.; Garbus, B. R.; Miller, W. G.; Williams, K. E.; Yee, E.; Bates, A. H.; Boyle, S.; Harden, L. A.; Cooley, M. B.; Mandrell, R. E. *Anal. Chem.* **2010**, *82*, 2717–2725.
13. Holland, R. D.; Duffy, C. R.; Rafii, F.; Sutherland, J. B.; Heinze, T. M.; Holder, C. L.; Voorhees, K. J.; Lay, J. O., Jr. *Anal. Chem.* **1999**, *71*, 3226–3230.
14. Mandrell, R. E.; Harden, L. A.; Horn, S. T.; Haddon, W. F.; Miller, W. G. Presented at American Society of Microbiology, Los Angeles, CA, 2000; Poster C-177.
15. Mazzeo, M. F.; Sorrentino, A.; Gaita, M.; Cacace, G.; Di Stasio, M.; Facchiano, A.; Comi, G.; Malorni, A.; Siciliano, R. A. *Appl. Environ. Microbiol.* **2006**, *72*, 1180–1189.

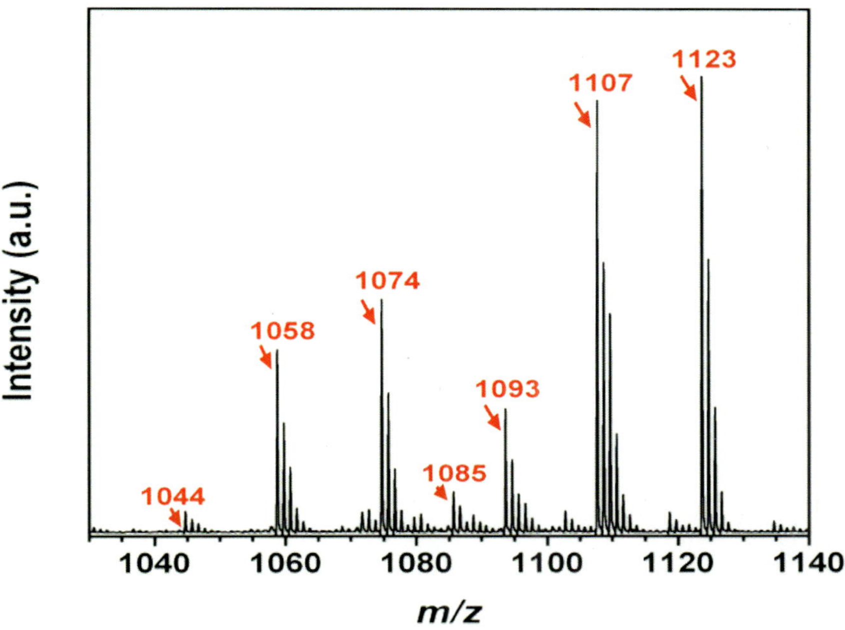

Figure 1b. Expanded view of B. subtilis ATCC 6633 lipopeptide products.

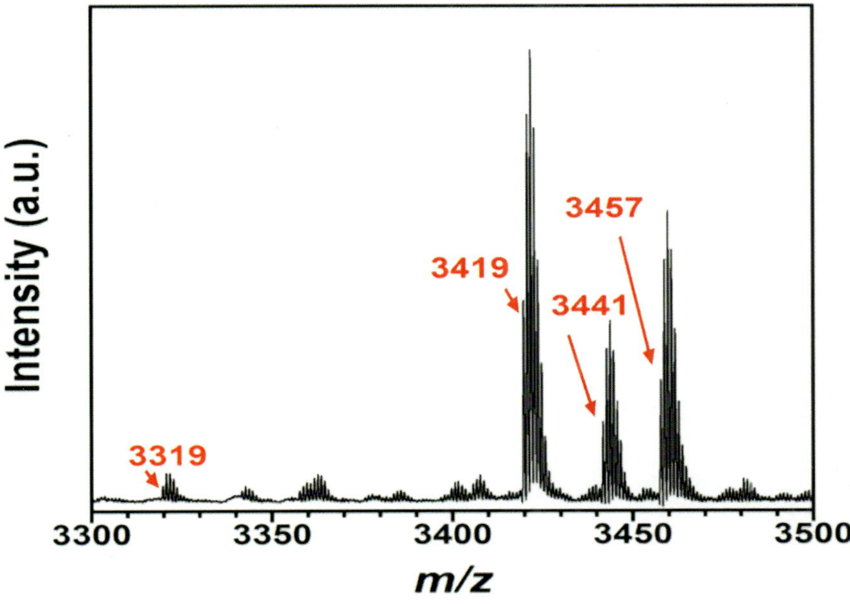

Figure 1c. Expanded view of B. subtilis ATCC 6633 subtilin with adducts and a protecting group.

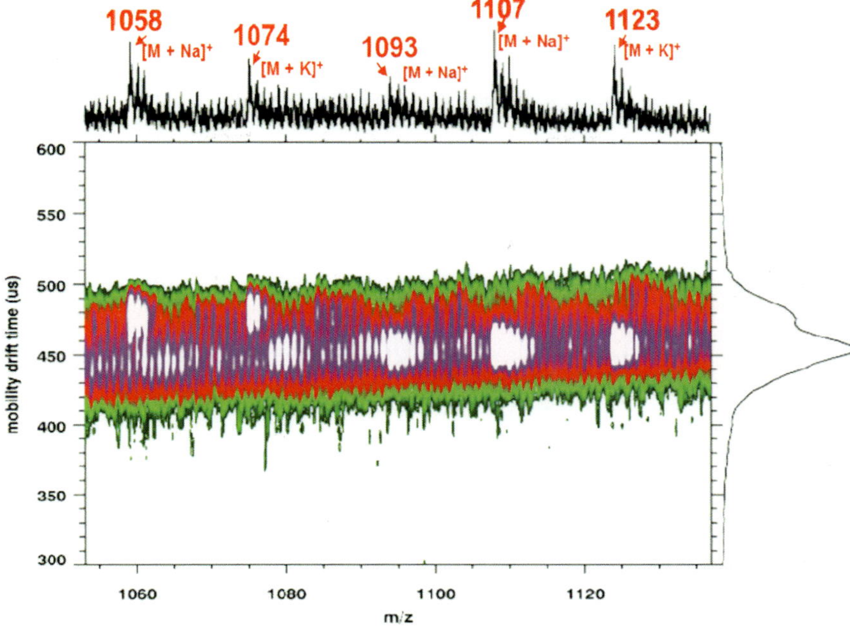

Figure 3. UV MALDI-IM-TOF MS 2-D contour plot of lipopeptide products with Na^+ and K^+ adducts from whole cell B. subtilis ATCC 6633.

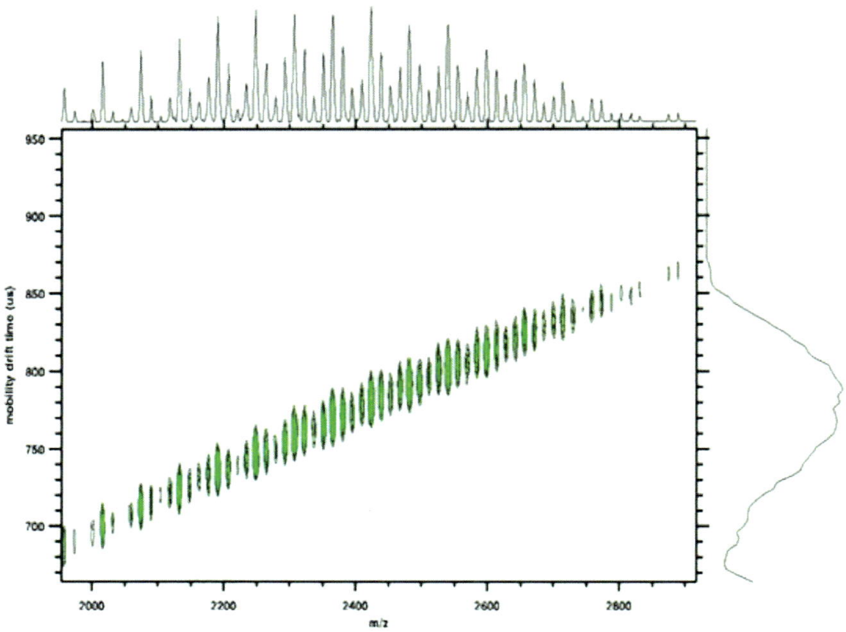

Figure 4. UV MALDI-IM-TOF MS 2-D contour plot of fatty acids separated by 14 Da from whole cell B. subtilis ATCC 6633.

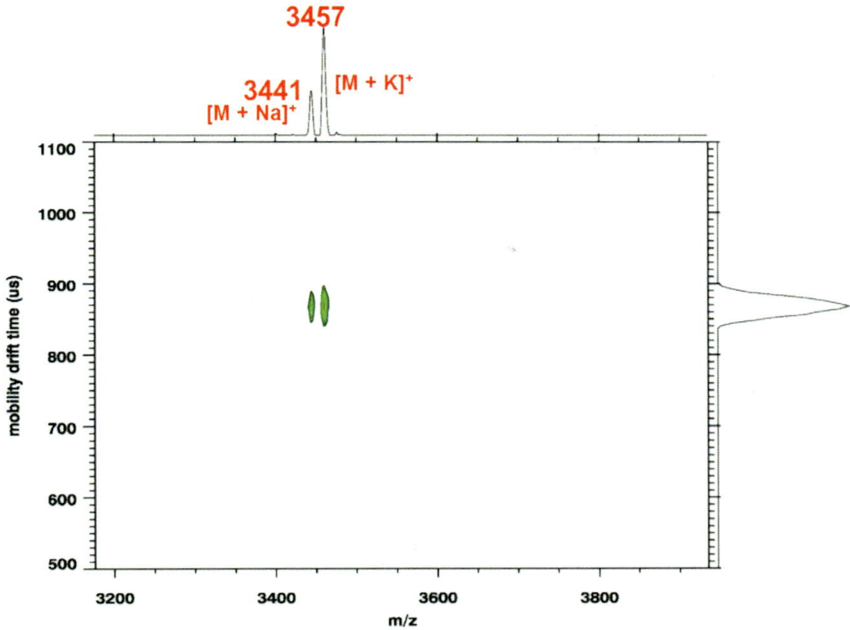

Figure 5. UV MALDI-IM-TOF MS 2-D contour plot of whole cell B. subtilis ATCC 6633. The peaks correspond to N-terminal succinylated subtilin with Na^+ *and* K^+ *adducts.*

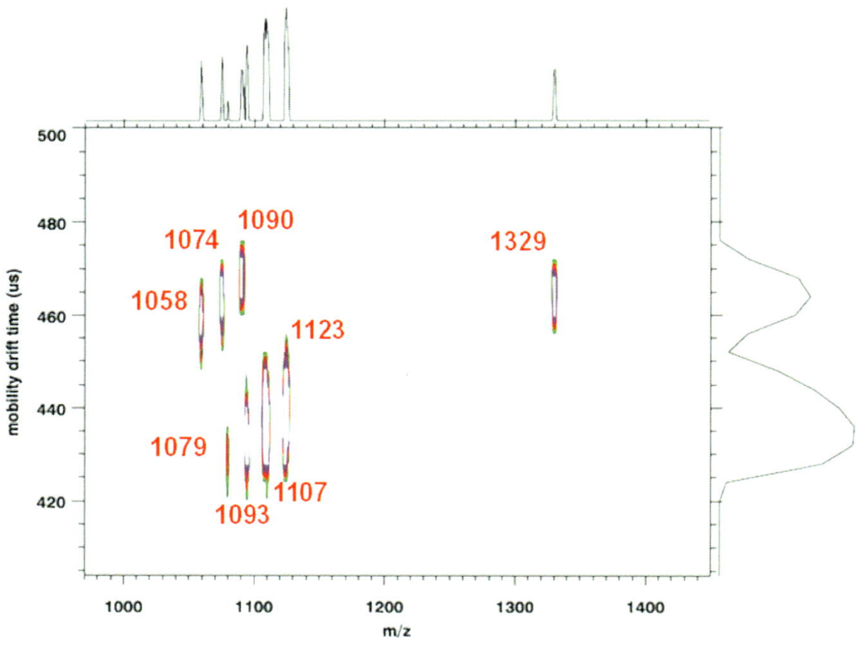

Figure 6. VUV post-ionization MALDI-IM-TOF MS 2-D contour plot of whole cell B. subtilis ATCC 6633. Additional peaks at 1079, 1090, and 1329 m/z were observed.

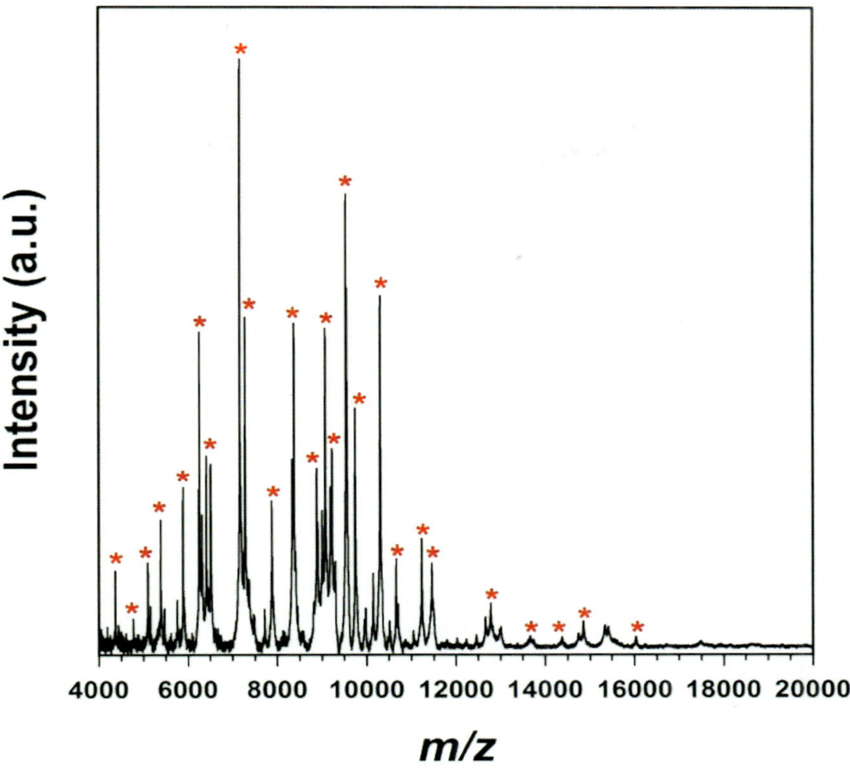

Figure 7. MALDI- TOF MS mass spectra of E. coli ATCC 9637 using α-cyano-4-hydroxycinnamic acid (CHCA).

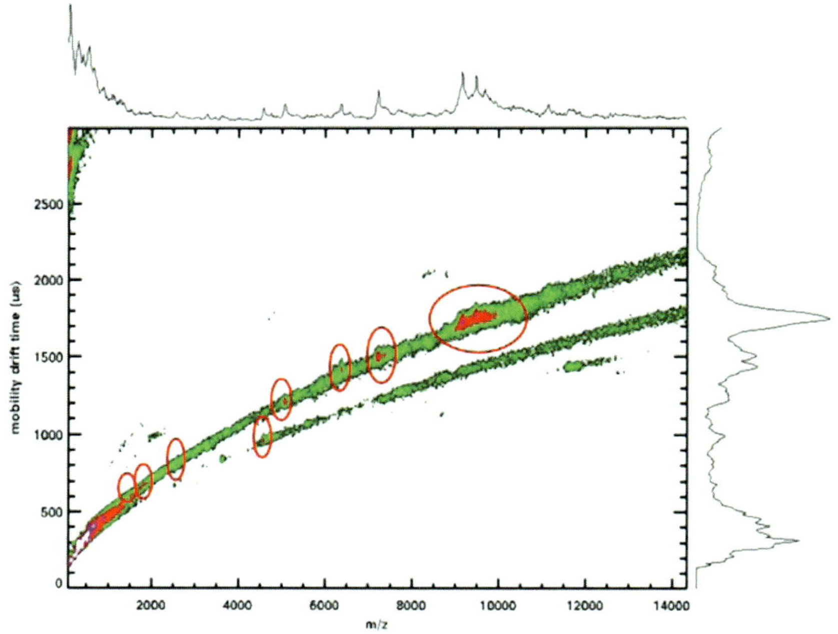

Figure 8. UV MALDI-IM-TOF MS 2-D contour plot of whole cell E. coli ATCC 9637.

Chapter 8

Top-Down Protein Analysis and Phylogenetic Characterization of Unsequenced Bacteria

Nathan J. Edwards,[*,1] Colin Wynne,[2] Avantika Dhabaria,[2] and Catherine Fenselau[2]

[1]Department of Biochemistry and Molecular & Cellular Biology, Georgetown University Medical Center, Washington, DC
[2]Department of Chemistry and Biochemistry, University of Maryland, College Park, MD
[*]Email: nje5@georgetown.edu

The identification of intact proteins by searching protein sequence databases is a powerful application of tandem mass spectrometry and bioinformatics. Applied to bacterial lysates, this 'top-down' strategy can identify proteins from microorganisms lacking annotated, sequenced genomes, and thereby establish sufficient protein sequence for phylogenetic characterization. Top-down analysis can also provide evidence for post-translational modifications and amino-acid substitutions in unsequenced microorganisms' proteins, and identify errors in protein annotations in closely related bacteria with a previously annotated genome sequence. The approach enjoys the advantages of mass spectrometry – speed, speed, sensitivity and specificity – and offers some unique analytical challenges.

Introduction

The characterization of microorganisms using proteomic technologies and bioinformatics offers greater flexibility in both establishing the experimental conditions and selecting the targets than do the various forms of mass spectral fingerprinting. A major limitation to the proteomics/bioinformatics approach would appear to be the absence of genomic sequences for some contemporary targets, and indeed for the vast majority of global prokaryote species. An approach is described in this chapter, which has been shown to allow some bacteria to be classified phylogeneticly, despite lacking sequenced genomes. Top-down analysis of proteins, based on high resolution tandem mass spectrometry, can provide reliable identifications of protein sequences, and thereby characterization of the bacterium. We present arguments here that information can also be deduced on amino-acid substitutions and post-translational modifications absent the genome sequence.

Experimental Considerations

The structural information provided by fragmentation of intact proteins requires activation of heavy ions. Collisional impact (CID) (*1*), electron attachment or transfer (*2*, *3*) and photoactivation (IRMPD) (*1*, *4*) have all been utilized successfully to induce fragmentation of intact proteins. Because of the additional energy provided by charge repulsion, all of these activation methods work most effectively on multiply charged ions, and these are provided most reliably by electrospray ionization. Electrospray ionization also allows for direct interface to HPLC. It is easy to see that the types of instruments that provide the most reliable top-down protein analysis are more complex than those used in mobile field systems (*5*, *6*), and it seems likely that top-down protein analysis will be carried out most successfully using stationary mass spectrometry systems in reference laboratories. Both precursor ions and product ions carry multiple charges in these experiments, which complicates the direct manual or automatic interpretation of fragment ions. In their initial exploration of top-down tandem mass spectrometry, McLafferty and co-workers laid out the requirement to decharge all product ions in order to facilitate and optimize interpretation (*1*). Identification of the charge state is readily made by analysis of carbon isotope clusters (*7*) however the resolution required to separate carbon isotopes limits the charge states that can be analyzed at high m/z values, and, consequently, the masses of the precursor and product ions that can be analyzed. Consequently, most applications have taken advantage of the high resolution offered by Fourier transform mass spectrometers.

In large scale applications of top-down protein analysis, one expects to analyze a complex mixture of proteins introduced from an HPLC column directly into the electrospray (or nanospray) ionization chamber. Computer controlled data-dependent tandem measurements made continuously throughout the chromatographic fractionation produce very large datasets, which must then be interpreted by bioinformatics software. Ideally, decharging, database searching, and spectral interpretation can be carried out without user intervention. Mass spectrometry and bioinformatics are rapid compared to chromatographic fractionation, and usually the chromatography limits the speed of a forensic analysis. However, chromatographic fractionation alleviates the problem of suppressed or selective ionization/desorption encountered when mixtures are analyzed by MALDI and electrospray, and will allow significantly higher numbers of proteins to be analyzed top-down in a single experiment.

A Brief History

The ground breaking work on top-down analysis of proteins using FTMS in McLafferty's laboratory (*1*) has subsequently been extended by other laboratories (see, for example, Kelleher (*8*), Smith (*9*), and Muddiman (*10*)). In particular, this second wave of research has addressed the development of effective bioinformatics programs (*11*), the incorporation of interfaced HPLC (*12–14*) and the extension of top-down strategies to high resolution TOF analyzers (*15*, *16*) and to orbitrap analyzers (*14*, *17*).

This chapter discusses the application of top-down analysis to identify proteins in whole cell lysates of microorganisms whose genomes have yet to be sequenced and for which there are few protein sequences available; the application of these protein sequences in phylogenetic analysis to place the unsequenced microorganism in context; the inference of amino-acid mutations in identified proteins; and the correction of genome annotations in related species that have been sequenced. The power of top-down MS/MS is reported by others in this book for detection of targeted biomarkers in regulatory applications, and it also offers great potential in hypothesis-driven studies of prokaryote (and eukaryote) biochemistry.

Top-Down Analysis of Bacterial Lysates by LC-MS/MS

One workflow for top-down analysis of whole cell lysates is illustrated in Figure 1 (*14*, *18*). Cultured or collected bacteria are lysed with 10% formic acid and centrifuged. The supernatant is desalted and injected into a protein-appropriate capillary HPLC column and the proteins are eluted with an acetonitrile gradient into a nanospray ionization source. Ions are activated by multiple low energy collisions in a linear ion trap analyzer interfaced to an orbitrap. Both precursor and product ions are analyzed at 15,000 to 60,000 resolution in the orbitrap. The size of interpretable proteins is limited by the resolution and the capability it provides to assign high charge states. ProSightPC 2.0 software (*11*) is used to decharge precursor and product ions and to match product ion spectra against b-ion and y-ion masses computed from protein sequences. FASTA format protein sequence databases containing a comprehensive set of related organisms' protein sequences are downloaded from the Rapid Microorganism Identification Database (RMIDb.org) (*19*). Identified proteins and their orthologs, determined using BlastP (*20*) on the RMIDb derived sequences, are assembled for phylogenetic analysis using phylogeny.fr (*21*), establishing the phylogenetic placement of unsequenced bacteria relative to closely related, sequenced species. Currently 10^4 lysed cells provide sufficient sample to identify 10 to 20 low mass proteins.

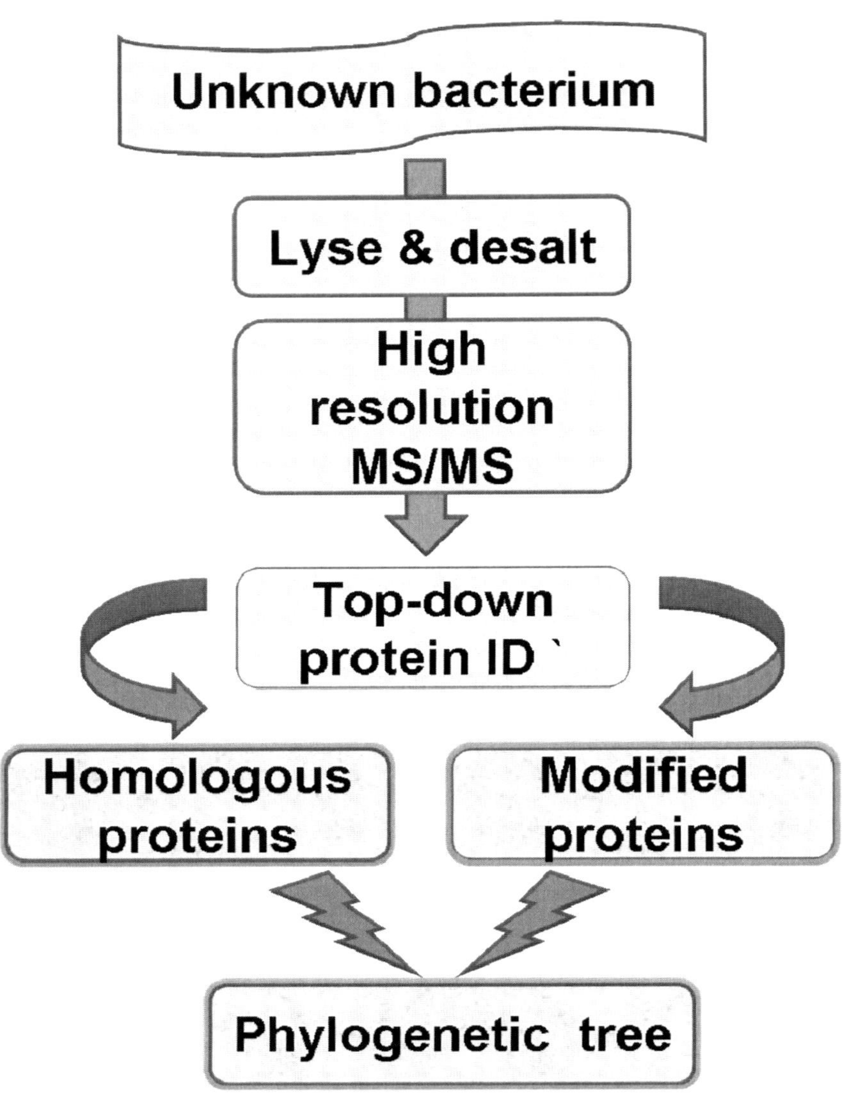

Figure 1. Workflow for top-down analysis of bacterial proteins and phylogenetic trees.

Protein Identification Informatics for Top-Down Spectra

There are considerable informatics challenges in the successful analysis of top-down intact protein fragmentation spectra. Much of the intuition derived from bottom-up protein identification informatics workflows can be applied, but top-down tandem mass-spectra have specific characteristics that require special care. We discuss the special requirements of top-down analysis next.

Spectral Processing

Spectral (pre-)processing of the acquired spectra is crucial for successfully interpreting CID top-down tandem mass-spectra. Intact protein precursors are often observed in high charge states, and consequently, so too the fragment ions in the corresponding MS/MS spectrum. Current sequence database search tools work poorly when the charge-states of fragment ions are undetermined and all possible charge-states must be enumerated. This is not a significant issue for interpreting bottom-up peptide fragmentation spectra, which have predominantly charge-state +1 fragment ions, but top-down spectra often contain fragment ions with a wide range of (high) charge states. As such, fragment ions must be de-charged to compute their uncharged mass. Similarly, the precursor mass must be determined.

In the Orbitrap, resolution degrades with the square-root of the m/z value (*22*). The maximum charge state resolvable at m/z value M can be computed by the formula:

$$z_{\max} = \left\lfloor \frac{20R}{M\sqrt{M}} \right\rfloor$$

where R is the resolution at 400 m/z. With a resolution setting of 30,000 @ 400 m/z, the Orbitrap can distinguish the isotope-cluster peaks of a charge-state +26 precursor ion at m/z value 800, and a charge-state +18 precursor at m/z value 1,200. Similarly, even with resolution setting 15,000 @ 400 m/z, fragment ion measurements can distinguish the isotope cluster peaks of a +20 fragment ion at m/z value 600 and a +13 fragment ion at m/z value 800. Measured with sufficient resolution in the survey and product ions scans, then, the precursor and fragment ions can be readily de-charged, monoisotopic masses determined, and faux, charge-state +1 spectra, without non-monoisotopic peaks, output for searching.

It is important to recognize, however, that spectral processing can introduce a variety of artifacts into these faux spectra. Common mistakes include the omission of fragmentation ions from low-intensity isotope-clusters; misalignment of precursor and fragment ion isotope clusters, resulting in ±1 and ±2 Da mass errors; and a limited ability to correctly de-convolute "intertwined" fragment ion isotope clusters or those with shared peaks. Off-by-one errors are more common for high-charge state precursor and fragment ions, as the monoisotopic peak disappears and the isotope cluster becomes flatter and more symmetric (*23*). Average masses can often be more reliably determined than monoisotopic masses, for high-charge state precursors, as average mass can be computed using only the observed, high-intensity peaks and their spacing.

A related issue is observed in the precursor m/z value recorded by the instrument. The mass-spectrometer records, and reports, a high-accuracy measurement of the peak selected for MS/MS – usually the most abundant peak of the precursor ion's isotope cluster. For intact protein precursors, the most abundant isotope cluster peak is not the monoisotopic peak, and is different from the average mass. The most abundant peak of the precursor's experimental isotope cluster may not even be the most abundant peak of the theoretical isotope cluster, due to variation in the observed peaks' ion current. The most abundant precursor isotope peak, computed *in silico* using averagine, is generally less than one Da from the average mass of the cluster, but the experimental most abundant precursor isotope cluster peak is sometimes as much as 2 Da further away from the average mass.

As a result of these issues, and despite the accurate measurement of the precursor ion's most abundant peak, we must generally allow the search engine to match precursor ions 2-3 Da from the theoretical value computed *in silico*, whether monoisotopic or average masses are considered.

Software available for processing high-resolution top-down spectra was developed originally for FT instruments but can be readily applied to Orbitrap data. Until the recent development of automated LC workflows for intact proteins, these tools would require the user to select and process each spectrum one-at-a-time. The automated acquisition of hundreds or thousands of top-down spectra renders this infeasible, so some tools now support spectral processing of all top-down spectra in a spectral data-file without user intervention.

Two such tools, THRASH (*7*) and Xtract (*24*), are integrated with the ProSightPC software (*11*). Version 2.0 of ProSightPC supports a High-Throughput mode to automate the spectral processing of an entire acquisitions' top-down spectra. THRASH requires profile spectra, while Xtract can work with either profile or peak-detected spectra. We have had good results from ProSightPC's implementation of TRASH, as implemented in the High-Throughput mode of ProSightPC version 2.0.

MS-Deconv (*25*), from the Pevzner lab, takes a different approach, taking peak detected spectra from ReAdW (Institute for Systems Biology, Seattle, WA) or msconvert (*26*) and deconvoluting the interleaved and shared peak isotope clusters to determine a de-charged monoisotopic peak list.

Protein Sequence Database

The second crucial element of successful top-down protein identification is the construction of an appropriate database of protein sequences in FASTA format. When the sample is known to come from a specific microorganism with a sequenced genome and a full complement of protein annotations are readily available, then these protein sequences can be downloaded from UniProt (uniprot.org) or NCBI (ncbi.nlm.nih.gov). The UniProt web-site search facility can be used to select complete proteomes from any species, strain, or other taxonomic grouping, which can then be downloaded in FASTA format. The NCBI ftp-site bacterial genome section contains the genome sequence and RefSeq protein sequence annotations for sequenced bacteria. Both sources provide an easy way to get access to protein sequences for specific species.

However, we have found that different sources of bacterial protein sequence for the same species are often inconsistent with each other, particularly when species are represented by many different strains. These inconsistencies are rarely due to genomic variation between strains, but are instead a consequence of different annotation software pipelines used. Even for a single well-characterized species, protein sequences from many sources should be searched – the Rapid

Microorganism Identification Database (RMIDb.org), described below, makes this possible.

When the sample being analyzed is not from a microorganism with a sequenced genome, our previous work (*14*) has shown that closely related, sequenced, bacteria often contain exactly conserved protein sequences. In this case, then, all protein sequences from a taxonomic subdivision that includes the sample should be assembled, at the genus, family, or order level. While UniProt makes it possible to select proteins according to taxonomic subdivision, in this case it is even more important to recruit relevant protein sequences from as many sources as possible, using the RMIDb.

The RMIDb (*19*) contains all bacterial, virus, plasmid, and environmental protein sequences from the Ventor Institute's Comprehensive Microbial Resource (CMR), UniProt's Swiss-Prot and TrEMBL, Genbank's Protein, and RefSeq's Protein resources. In addition, Glimmer3 (*27*), a widely used bacterial gene-annotation program modified to generate alternative translation start-sites (*28*), is used to predict an aggressive set of putative proteins, including alternative translation start-sites, on all RefSeq bacterial genomes. RMIDb "models" representing a specific species or taxonomic subdivision will include all protein sequences from any of these sources, plus the aggressive protein predictions of Glimmer3, reducing the chance that observed proteins' sequences are omitted from the sequence database. The putative Glimmer3 protein sequences can also provide positive evidence for missing or erroneous protein annotations, when Glimmer3 protein sequences are the only match to a spectrum. RMIDb models can be downloaded as FASTA files from rmidb.org.

In addition to an exhaustive set of protein sequences, it is important to consider how N-terminal Met excision (NME) should be handled. While some search engines (notably ProSightPC) will automatically search both the cleaved and uncleaved protein sequences, other tools cannot. NME sequences can be readily enumerated, at most doubling the size of the sequence database. A more careful enumeration, which takes into account NME cleavage potential (*29*), will increase the size of the sequence database by about 25%. Other post-translational modifications, such as N-terminal acetylation, are best handled by the search engine. RMIDb models can be configured to enumerate none, likely, or all NME protein isoforms. In our recent work (*14*, *18*), we selected all *Enterobacteriaceae* genus protein sequences using the RMIDb and enumerated likely NME protein forms.

Finally, the sequence database must be configured for the search engine. Details depend on the specific search engine employed, but ProSightPC 2.0, in particular, enumerates NME and N-terminal acetylation isoforms as the FASTA database is loaded. Other search engines generally consider post-translational modifications only in the search phase.

Sequence Database Search with Top-Down Tandem Mass-Spectra

While there are few readily available tools designed specifically for searching protein sequence databases with intact protein fragmentation spectra – ProSightPC is a notable exception – many tools designed originally for peptide identification from bottom-up tandem mass-spectra of peptides work quite well for top-down MS/MS spectra. In addition to ProSightPC, we have successfully applied Mascot, X!Tandem, and OMSSA to identify proteins using the faux charge-state +1 spectra resulting from spectral preprocessing by THRASH, implemented inside ProSightPC.

For search engines designed for bottom-up peptide identification workflows, it is necessary to turn-off *in silico* protein digestion. Some tools have a no-digest "enzyme", while others will allow you to configure impossible or rare cleavage motifs. Non-specific cleavage will permit any truncation of a protein sequence's N- and C-terminus to be compared against the top-down spectra. ProSightPC offers an absolute mass search option, equivalent to the no-digest "enzyme", and a biomarker search option, equivalent to a non-specific cleavage search.

Regardless of the search engine, precursor ion matches will generally require at least a 2 Da deviation from the theoretical monoisotopic or average masses, as previously discussed. Fragment ions should be matched with relatively tight mass-tolerances – such as 15 ppm. While some fragments may be missed due to incorrect monoisotopic peak determination, the specificity of the other fragmentation ion matches is more important. For CID top-down spectra, the search engines match b- and y-ions with the fragment ions, just as for bottom-up peptide identification workflows.

The statistical significance of protein identifications from top-down analyses can be difficult to establish, as research into appropriate techniques is still rather immature. While most tools provide an E-value with each protein identification, it is important to realize that scores and E-value estimation techniques designed for bottom-up peptide CID spectra may be poorly suited to high-resolution fragment ion measurements and sparse top-down CID fragmentation. Decoy sequence databases can be constructed by reversing or shuffling the target sequence database, as for bottom-up peptide identifications, but decoy-based FDR is a crude metric at best, and has not be thoroughly explored for top-down protein identifications. Given the immaturity of the statistical significance models for accurate fragment ion MS/MS spectra of intact proteins, we advise caution in choosing statistical significance cutoffs. ProSightPC suggests a 10^{-4} E-value threshold, by default, for example.

Figure 2 shows an example of an intact protein CID tandem-mass-spectrum (precursor m/z value 596.96, in charge-state +16) observed in an *E. coli* top-down analysis. ProSightPC 2.0 was used to search *Enterobacteriaceae* protein sequences and DNA-binding protein HU-alpha was identified, with *E*-value 2e-22, based on 11 b-ion and 14 y-ion matches. We point out the high-charge state fragment ions, which must be decharged using spectral processing.

Figure 3 shows another intact protein CID tandem-mass-spectrum (precursor m/z value 1014.30, in charge-state +8) observed in an *E. herbicola* top-down analysis. *E. herbicola* genome has yet to be sequenced. ProSightPC 2.0 was used to search *Enterobacteriaceae* protein sequences and translation initiation factor 1A from *E. tasmaniensis* and other species, was identified, with *E*-value 2.34e-28, based on 10 b-ion and 16 y-ion matches.

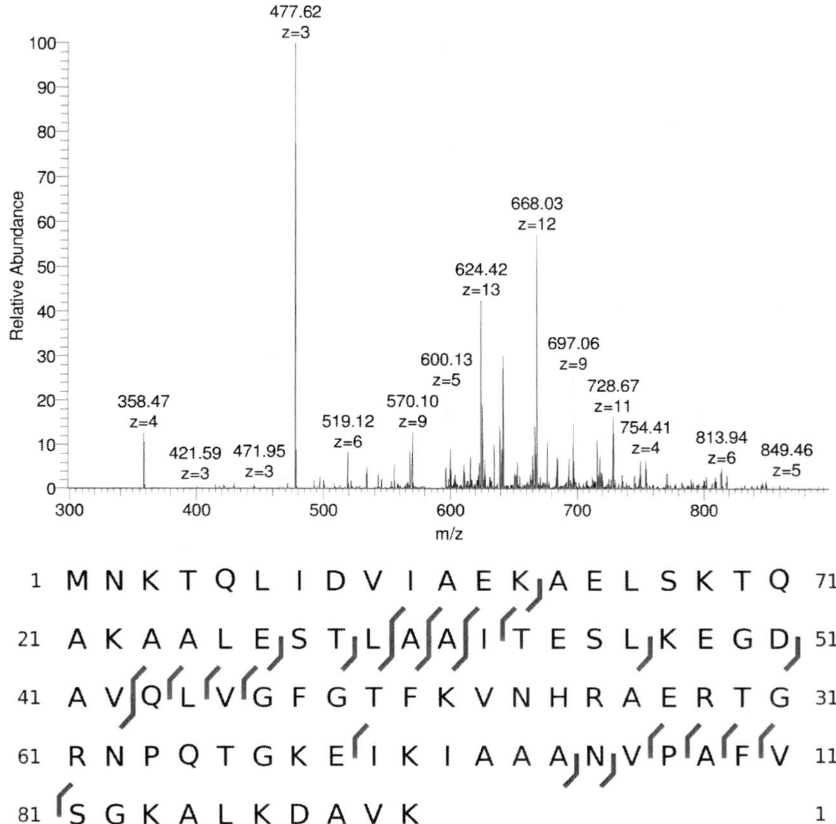

Figure 2. Product ion mass spectrum (top) from an intact protein from E.coli with a precursor ion at m/z 596.96 in charge-state +16. The protein was identified as DNA-binding protein HU-alpha with E-value 2e-22, based on 25 b- and y-ion fragment matches (bottom).

Beyond Protein Identification

With the capability in hand for protein identification from top-down tandem mass-spectra from bacteria lacking sequenced genomes, we can explore techniques for recognizing, and characterizing, amino-acid sequence mutations and post-translational modifications; methods for characterizing unsequenced bacteria by carrying out phylogenetic analysis using identified proteins; and the correction of genomic annotations in related organisms using identified proteins.

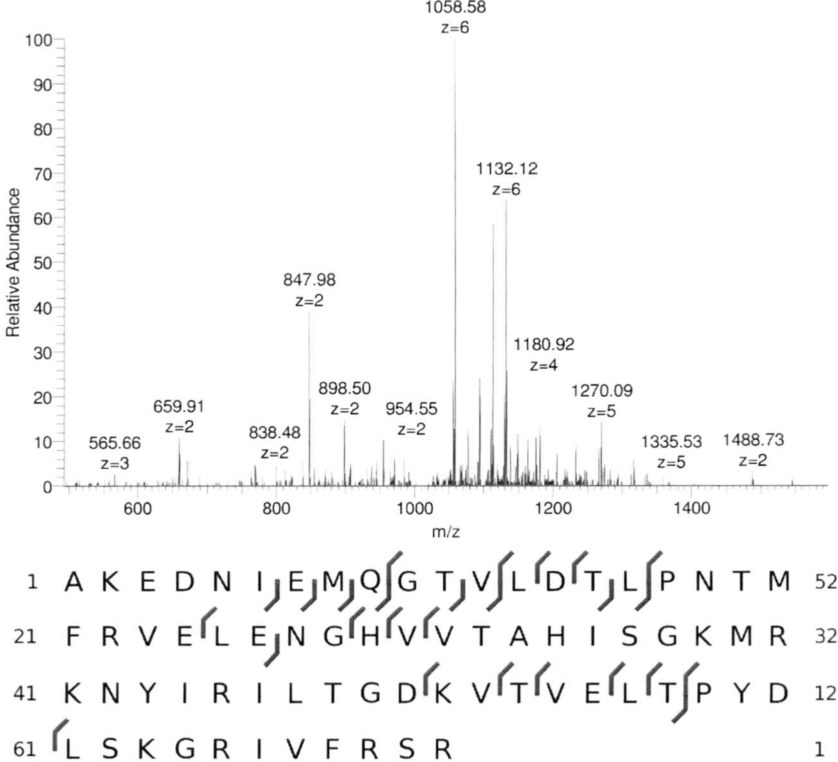

Figure 3. Product ion mass spectrum (top) of an intact protein precursor ion from E. herbicola with a precursor ion at m/z 1014.30 in charge-state +8. The protein was matched to translation initiation factor 1A from E. tasmaniensis with E-value 2.34e-28 based on 26 b- and y-ion fragment matches (bottom). Adapted from (18).

Mutation and Modification Tolerant Sequence Database Search

A significant concern with sequence database searches using top-down protein fragmentation spectra is that even a single amino-acid substitution or post-translational modification will change the protein's molecular weight, making a match between the experimental and theoretical precursor mass impossible. While this issue is also a concern for sequence database searches of peptide fragmentation spectra, tryptic peptides are short which reduces the chance of an unexpected mutation or modification. These issues are exacerbated when searching homologous protein sequences from related organisms, since amino-acid mutations in orthologous proteins are quite common, even for closely related species. Indeed, it is remarkable that we are able to match any top-down spectra to protein sequences from related species, a reflection of the (very) high degree of conservation for ribosomal proteins and the density of related *Enterobacteriaceae* species with sequenced and annotated genomes.

Nevertheless, we observe, as have others (*1*, *30*, *31*), that the number and specificity of high-resolution fragment ion matches makes matching the precursor ion accurately unnecessary to identify the protein with high specificity. Specifically, we use a 250Da precursor match tolerance in order to consider spectrum-protein matches that may be different by a small number of mutations or post-translational modifications. When the search engines score these potential matches, b- and y-ions are computed directly from the amino-acid sequence, and have the correct mass until the changed residue(s) are incorporated in the theoretical fragment mass computation. Crudely, half of the b- and y-ions will still match their theoretical mass, despite an unmodeled difference between the experimental precursor and the theoretical protein sequence. If the changed residue is at the N-terminal, only b-ions will be lost, while if the changed residue is at the C-terminal, only y-ions will be lost. For an unexpected mass-shift in the middle of the protein, all b-ions and y-ions up to the changed residue will still be matched. When these partial matches are still sufficient to achieve a good score, we observe these spectrum-protein matches as significant protein identifications.

Figure 4 shows an example of a mutation tolerant match to a homologous protein sequence for an *E. herbicola* intact protein tandem mass spectrum (precursor m/z value 732.71, in charge state +13) which matches DNA binding protein HU-alpha from *E. coli* and other species, (cf. Figure 2) with E-value 7.50e-26 and 31 b- and y-ion fragment matches, and a precursor mass-delta of −14.13. The b- and y-ion matches stop at amino-acids 37 and 42, respectively, due to an unknown mass-shift or mass-shift(s) in the protein. Below, we discuss methods for interpreting these mutation tolerant matches.

We point out that when the protein is matched with a non-trivial number of b- and y-ions that a potentially large portion of the N- and C-terminus of the protein sequence is confirmed, as the fragment matches establish that no mass-shifts have occurred. More care must be taken when only b- or only y-ions are matched, as the sequence of only one terminus is established. If b- and y-ion matches cross then the full length protein sequence is established, and the absence of any mass-shift can be confirmed using the precursor mass. In a top-down analysis of *E. herbicola* proteins using this technique, we considered the N-or C-terminal sequence confidently established if confirmed by at least 3 fragment ion matches. Figure 5 summarizes the confidently established amino-acid sequences of a set of proteins identified from *E. herbicola.* Eight of the 14 proteins were identified despite unexpected mass-shifts.

Phylogenetic Characterization Using Top-Down Protein Identifications

Presented with an unknown microorganism, top-down protein identifications can often be used to suggest the organism's identity, when all identified proteins point to a single species (*15*). However, if the unknown microorganism has not yet been sequenced, we may identify proteins from a hodgepodge of related species. Figure 6 shows how proteins identified in a top-down analysis of *Y. rohdei* (*14*) are found in a complex pattern of related species. Phylogenetic analysis based on top-down protein identifications permits a nuanced determination of the appropriate taxonomic placement of the unknown microorganism.

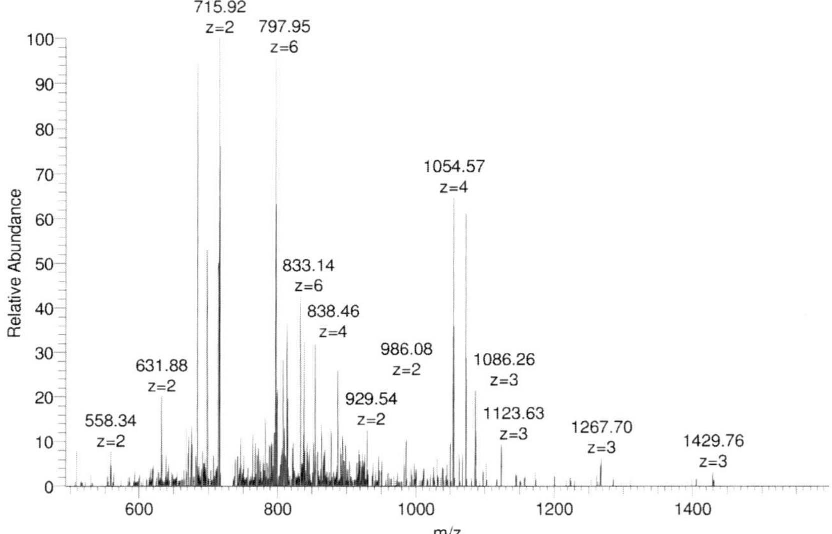

1 M N K T Q L I D V I A D K A E L S K T Q 72
21 A K A A L E S T L A A I T E S L K E G D 52
41 A V Q L V G F G T F K V N H R A E A R T 32
61 G R N P Q T G K E I K I A A A N V P A F 12
81 V S G K A L K D A V K 1

1 M N K T Q L I D V I A D K A E L S K T Q 72
21 A K A A L E S T L A A I T E S L K D G D 52
41 A V Q L V G F G T F K V N H R A E A R T 32
61 G R N P Q T G K E I K I A A A N V P A F 12
81 V S G K A L K D A V K 1

Figure 4. Product ion mass spectrum (top) of an intact protein precursor ion from E. herbicola at m/z 732.71 in charge-state +13. The protein was indentified as DNA binding protein HU-alpha with E-value 7.50e-26 based on 31 b- and y-ion fragment matches, with a precursor mass-delta of -14.13 (middle). After substitution of D for E (mass-delta -14) at the 38th position (highlighted), the E-value becomes 1.91e-58, with 41 b- and y-ion fragment matches (bottom). Adapted from (18).

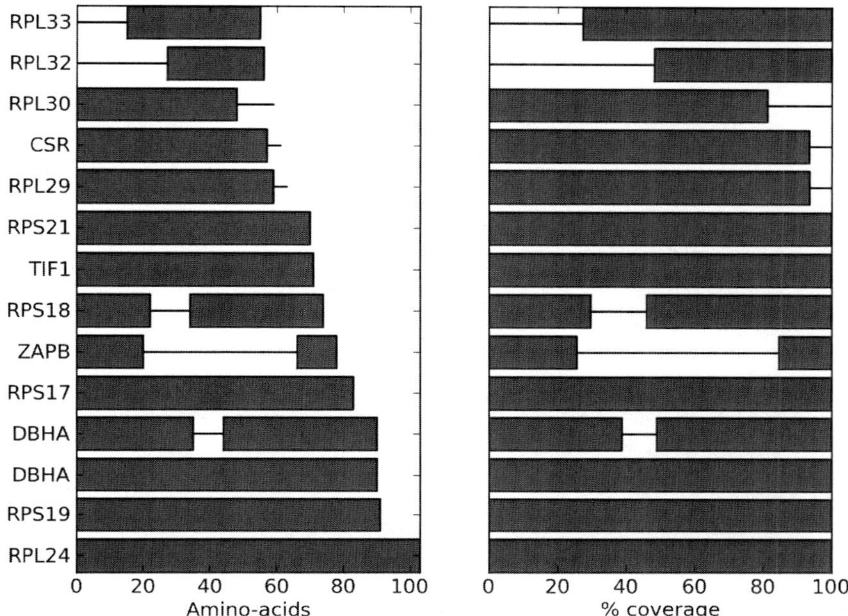

Figure 5. Representation of confidently established N- and C-terminal sequences in a top-down analysis of E. herbicola using a mutation tolerant search strategy.

Phylogenetic analysis proposes an evolutionary explanation in the form of a phylogeny or tree of mutations sufficient to explain the observed amino-acid or nucleotide changes between (known) orthologous genes or proteins from related modern-day organisms. Phylogeny analysis is generally carried out on the basis of an underlying mathematical or probabilistic model of evolution. See Mount (*32*) for a gentle introduction.

In order to characterize an unknown microorganism using a phylogeny analysis, we require the sequence of a protein from the organism and a (large) cluster of orthologous proteins in known species. The orthologous proteins are analyzed using a multiple sequence alignment, which aligns conserved amino-acids and establishes the amino-acid positions with mutations in at least one sequence. Amino-acid positions with mutations form the input to the tree-building phase of the phylogeny analysis.

Phylogenies computed on the basis of one protein's orthologs will not necessarily agree with those computed from another protein's orthologs. Proteins evolve at different rates, and specific loci in each protein may be under stringent functional constraint. Furthermore, ortholog clusters may be small or large, depending on the availability of protein sequences, degree of conservation, and the evolutionary age of the protein.

A single "average" phylogeny may be obtained by concatenating the protein sequences from each species. To guard against artifacts created by the order in which protein sequences are joined, we fix a random permutation of the proteins, join each species' sequences, and then conduct a multiple sequence

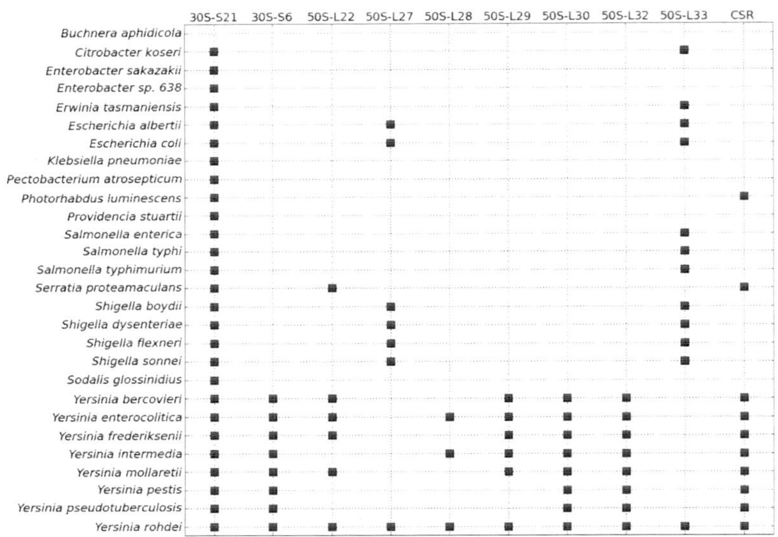

Figure 6. Incidence matrix for observed Yersinia rohdei proteins matched in Enterobacteriaceae species. Boxes indicate an exact sequence match. (Reprinted from reference (14), with permission from the American Chemical Society)

alignment and phylogenetic analysis. We use the web-site phylogeny.fr (*21*) to conduct the phylogeny analysis and draw the phylogenetic trees. Figure 7 shows the phylogenetic placement of *Y. rohdei* (*14*) in the context of other *Enterobacteriaceae* species.

Given protein identifications by top-down tandem mass-spectra, computed as above, it is straightforward, though heuristic and tedious, to establish orthologs for each identified protein. We use BlastP (*20*) to find homologous proteins in the *Enterobacteriaceae* FASTA database and select the one with the smallest E-value for each species. When only partial N- and C-terminal sequence is confidently established, as in Figure 5, we must only consider sequences that align with the confidently established amino-acids in the identified proteins. Some amino-acid positions which have undergone mutations may be lost, but enough sequence is established for a successful phylogeny analysis.

Mutation and Modification Inference from Top-Down Protein Identifications

As already discussed, mutation and modification tolerant sequence database searches can be used to identify proteins even though they have undergone a small number of changes. Of particular interest are cases in which these changes can be located accurately on appropriate residues of the identified protein's sequence. A typical scenario is shown in Figure 4, where multiple b-ions match from the N-terminal, and end at residue 37, and multiple y-ions match from the C-terminal, and end at residue 42. The precursor mass-delta suggests that a mass shift of −14.13 Da should be placed somewhere between the b- and y-ion matches. Placement of

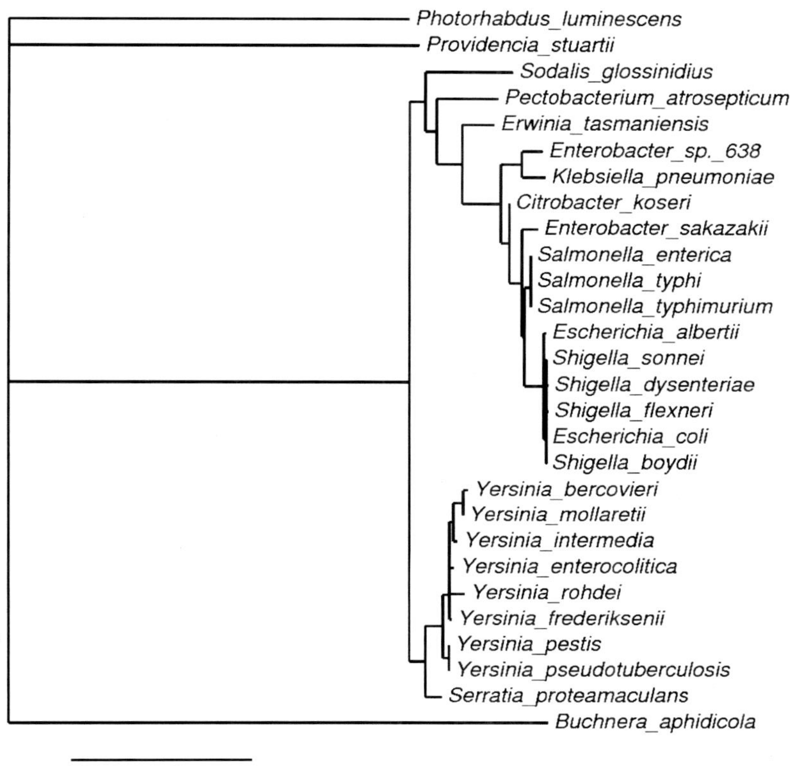

Figure 7. Phylogenetic tree for Y.rohdei based on top-down protein identifications. Adapted from (14).

this mass-shift on the Asp residue in position 38 improves the E-value and induces an additional 10 b- and y-ion matches across the protein sequence.

While the putative placement of a single mass-shift can be readily accomplished by checking possible placements in turn, some computational assistance may be necessary to evaluate E-values and other scores. ProSightPC's Sequence Gazer recomputes E-values and scores for user-specified mass-shifts on ProSightPC identifications. Once placed, a mass-shift may result in further b- and y-ion fragment matches that may further constrain the putative placement of the mutation or modification still further.

A related approach, proposed by the Pevzner group (*33*), uses dynamic programming based spectral alignment to evaluate all possible placements of expected and arbitrary mass-shifts on the amino-acid sequence. Unfortunately, the sparse fragmentation of CID top-down spectra and spurious fragment matches to low-abundance ions make such general mass-shift placements somewhat speculative. Nevertheless, this approach can locate multiple mass-shifts on the amino-acid sequence, which would be impossible in a manual analysis.

This simple discussion, however, masks some important issues. One and two Dalton errors in inferred precursor masses are common, suggesting mass-shifts one or two Daltons from the correct value, and making +15 indistinguishable, without further careful analysis, from +14 and +16. Analogously, precursor mass-deltas of one or two Daltons are much more likely to be explained by a misassigned monoisotopic precursor peak, rather than one or two Dalton amino-acid substitutions.

Another issue, which can occur when the sequence database and/or search strategy enumerates equivalent protein variants, occurs when the same identification is reported in two different ways. For example, the identification of a protein (initial Met retained) with a −131 precursor mass-delta is an isobaric equivalent to the identification of the NME variant of the same protein with a small precursor mass-delta, and cannot be distinguished if there are no b-ion matches. Recognizing these events is made even more difficult by the off-by-one precursor mass errors already discussed. The biological interpretation is the same in either case, but search tools may report one or both of these identifications. After all, the theoretical fragment masses computed from these two alternatives are the same.

Common mass-deltas include −131 (NME), −89 (NME + acetylation), +14 (methylation), +16 (oxidation), +28 (two methylations), +42 (acetylation), and their off-by-one equivalents. Negative mass-deltas are difficult to explain as PTMs, and appropriate amino-acid mutations should be considered, particularly when identifying proteins across species, as with our homology based strategy. Table 1 lists the (integerized) mass-deltas explained by all amino-acid substitutions that can be made with a *single* nucleotide change. Substitutions for which no pair of corresponding codons differ by one nucleotide are not shown. Notice that there are multiple possibilities for many mass-deltas – even when considering a small number of amino-acid positions. While the substitution of Figure 4 (E38 → D) is consistent with the precursor mass-delta, so too is the isobaric substitution A41 → G, with only a few b-ions to help determine the most reasonable placement.

The difficulty with placing mass-shifts appropriately becomes apparent as the number of possible explanations for each mass-shift grows. The PSI-MOD project (*34*) seeks to create a modern ontology for biological and chemical modifications to succeed RESID, Unimod, and deltamass, and has been projected to ultimately list as many as 20,000 modifications (*35*), significantly more than the approximately 500 modifications in RESID. It is important to remember that the objective here is to provide a biological or chemical explanation for observed mass-shifts, something that is not provided in the top-down mass-spectrum. Where available, UniProt modification features may provide some clarity, as these capture the specific modifications known to occur on a particular protein. Nevertheless, these modification annotations are not well-integrated with search engines and tools for querying UniProt for these modifications are rudimentary, at best. We hope these tools will improve over time.

Table 1. Mass difference (in Daltons) for all amino-acid substitutions possible via a single nucleotide mutation. Substitutions requiring more than one nucleotide change not shown

	W	*Y*	*R*	*F*	*H*	*M*	*E*	*K*	*Q*	*D*	*N*	*L*	*I*	*C*	*T*	*V*	*P*	*S*	*A*	*G*
G	+129		+99				+72			+58				+46		+42		+30	+14	
A							+58			+44					+30	+28	+26	+16		−14
S	+99	+76	+69	+60							+27	+26	+26	+16	+14		+10		−16	−30
P			+59		+40				+31			+16			+4			−10	−26	
V				+48		+32	+30			+16		+14	+14						−28	−42
T			+55			+30		+27			+13		+12				−4	−14	−30	
C	+83	+60	+53	+44														−16		−46
I			+43	+34		+18		+15			+1	0			−12	−14		−26		
L	+73		+43	+34	+24	+18			+15				0			−14	−16	−26		
N		+49			+23			+14		+1			−1		−13			−27		
D		+48			+22		+14				−1					−16			−44	−58
Q			+28		+9		+1	0				−15					−31			
K			+28			+3	+1		0		−14		−15		−27					
E								−1	−1	−14						−30			−58	−72
M			+25					−3				−18	−18		−30	−32				
H		+26	+19						−9	−22	−23	−24					−40			

Continued on next page.

Table 1. (Continued). Mass difference (in Daltons) for all amino-acid substitutions possible via a single nucleotide mutation. Substitutions requiring more than one nucleotide change not shown

	W	*Y*	*R*	*F*	*H*	*M*	*E*	*K*	*Q*	*D*	*N*	*L*	*I*	*C*	*T*	*V*	*P*	*S*	*A*	*G*
F		+16										−34	−34	−44		−48		−60		
R	+30				−19	−25		−28	−28			−43	−43	−53	−55		−59	−69		−99
Y				−16	−26					−48	−49			−60				−76		
W			−30									−73		−83				−99		−129

Identification of Genome Annotation Errors Using Top-Down Protein Analysis

The use of peptide identifications to inform and correct, in particular, bacterial genome annotations has seen considerable attention in recent years. High-quality peptide identifications can suggest novel coding regions, erroneous N-terminal translation start sites, occurrence of NME or lack of NME, and amino-acid mutation. Top-down protein identifications offer a powerful advantage over bottom-up peptides in this context – the N- and C-terminal sequence of the protein can be readily determined in the majority of top-down identifications. For bacteria, which of course have no introns, establishing the N-terminal sequence, in particular, is essentially sufficient experimental evidence to establish the correct gene and protein annotation in the corresponding ORF.

In this context, the use of an aggressively inclusive *Enterobacteriaceae* protein sequence database which includes alternative translation start-sites is extremely important, as this makes it possible to identify proteins that otherwise would be missed. Furthermore, the incorporation of an exhaustive list of protein predictions from a variety of sources, species, and strains, from closely related organisms, makes it possible to assert the correct protein annotations in species other than the one under study.

Discrepancies in the protein annotations from different sources can be readily detected by pairwise homology searches, but without experimental data, there is no way to determine the correct annotation. These anomalies are quite easy to find and are usually computational artifacts, a consequence of the different software packages used to predict and annotate genes and proteins on bacterial genomes. These software pipelines often use homology to check or correct protein annotations, but this also propagates erroneous protein annotations made in annotating the first few genomes to be sequenced. Furthermore, available homologs may not be high-quality or from a particularly closely related species. While it is not unlikely that distantly related microorganisms might use different translation start-sites on a homologous gene sequence, we certainly expect that closely related organisms with highly conserved gene sequence will produce the same (up to homology) protein product. With bone-fide evidence of the N-terminus sequence of the protein derived from a top-down protein identification, we can confidently assert which of the many orthologous protein annotations in related species are correct, and which need correction.

We point out that we are presuming that each locus is responsible for the production of a single protein product. While we cannot be certain that these bacteria are not producing multiple protein isoforms from the same locus, in the absence of experimental evidence for two isoforms, it is not unreasonable to presume that the version supported by experimental evidence is the correct one.

We demonstrate these issues with 50S Ribosomal Protein L32, identified in *Y. rohdei* by virtue of identical sequence orthologs in *Y. enterocolitica*, *Y. pestis*, and *Y. pseudotuberculosis*, all from Swiss-Prot. None of the other Yersinia species with proteins matching the RPL32 protein family, *Y. aldovae, Y. bercovieri, Y. frederiksenii, Y. intermedia, Y. kristensenii, Y. mollaretii, Y. rohdei*, or *Y. ruckeri* contained the correct sequence, despite the integration of Genbank, RefSeq, TrEMBL, the Venter Institute's (JCVI) Comprehensive Microbial Resource (CMR), and Swiss-Prot protein sequence repositories in the RMIDb. Each of these species annotates a shorter version of the same protein sequence, choosing to denote the translation start-site 23 amino-acids (annotation variant V_{+23}) into the correct sequence. The erroneous protein sequence is provided by TrEMBL and RefSeq, for these species, whose genomic sequence is not available from NCBI. Interestingly, at the time of our interrogation the CMR provided the incorrect protein sequence as the JCVI (re-)annotation of *Y. enterocolitica*, and the correct protein sequence (from Swiss-Prot) as the original (non-JCVI) annotation. In the *Enterobacter* species, another sequence variant appeared, 13 amino-acids (annotation variant V_{+13}) into the correct sequence, for *E. cancerogenus* (CMR, TrEMBL), *E. cloacae* (Genome, TrEMBL), and *Enterobacter sp.* (JCVI re-annotation).

Differences in the annotated RPL32 protein sequences in the many available *E. coli* strains further demonstrate the issue. The JCVI re-annotation consistently selects variant V_{+13} for the *E. coli* strains, including strain *APEC O1*, only available from the CMR, which has no other annotation for RPL32. For strain *SMS-3-5* only, the JCVI annotation matches that of the external (correct) sources. TrEMBL and RefSeq do better on these sequences too, with only strain *83972* being annotated with the V_{+13} variant. RefSeq introduces a fourth sequence variant with start-site seven amino-acids upstream (annotation variant V_{-7}) of the experimentally observed start-site in strain *O157:H7 str. EC4206*. We point out that corrections can be made to these mis-annotations on the basis of just one of the many top-down identifications from (*14*, *18*, *25*).

Furthermore, we can establish completely missing annotations using the methodology – as the Glimmer3 annotations permit us to verify that the sequence is present in the genome under consideration. *Y. pestis Z176003* is one such organism, missing annotations in RefSeq and TrEMBL for ribosomal proteins S21 and S19 identified using *E. herbicola*, which are none-the-less output by Glimmer as a plausible ORF and translation start site. A protein sequence corresponding to DNA-binding protein was matched with and without its initial Met, top-down in *E. cloacae*, to a protein sequence representing a start-site missing from all traditional sources, and only matched due to the aggressive incorporation of Glimmer3 alternative start-site predictions in the RMIDb resource and the search substrate for ProSightPC.

Conclusion

- Unbiased, discovery workflow is established for acquisition of high-accuracy fragment ion top-down tandem mass-spectra from crude cell lysates.
- Top-down spectra are identifiable, even for proteins from microorganisms with little sequence in public repositories and those without sequenced genomes.
- Mutation and modification tolerance search strategies can boost the number of identifications, especially for cross-species searches.

- High-sequence coverage obtained from top-down protein identifications can be used to conduct phylogeny analyses, making it possible to place organisms without DNA sequence in their correct phylogeny.
- Careful consideration of fragment ion matches and reasonable mass-shift explanations can sometimes suggest plausible explanations for the observed (but unknown) protein isoforms.
- Experimental verification, especially of N-terminal sequences of bacterial proteins, coupled with homology can establish the correct protein annotation variant for identified proteins, resolving the otherwise ambiguous determination of which annotation variant is correct.

References

1. Mortz, E.; O'Connor, P.; Roepstorff, P.; Kelleher, N.; Wood, T.; McLafferty, F. W.; Mann, M. *Proc. Natl. Acad. Sci. U.S.A.* **1996**, *93*, 8264–8267.
2. Zubarev, R. A.; Kelleher, N. L.; McLafferty, F. W. *J. Am. Chem. Soc.* **1998**, *120*, 3265–3266.
3. Chi, A.; Bai, D. L.; Geer, L. Y.; Shabanowitz, J; Hunt, D. F. *Int. J. Mass Spectrom.* **2006**, *259*, 197–203.
4. Madsen, J.; Gardner, M.; Smith, S.; Ledvina, A.; Coon, J.; Schwartz, J.; Stafford, G; Brodbelt, J. *Anal. Chem.* **2009**, *81*, 8677–8686.
5. Ecelberger, S. A.; Cornish, T. J.; Collins, B. F.; Lewis, D. L.; Bryden, W. A. *Johns Hopkins APL Tech. Dig.* **2004**, *25*, 14–19.
6. Sundaram, A. K.; Gudlavalleti, S. K.; Oktem, B.; Razumovskaya, J.; Gamage, C. M.; Serino, R. M.; Doroshenko, V. M. *Proceedings of the 56th Conference of the American Society for Mass Spectrometry*, Denver, CO, June 1−5, 2008.
7. Horn, D. M.; Zubarev, R. A.; McLafferty, F. W. *J. Am. Soc. Mass Spectrom.* **2000**, *11*, 320–332.
8. Kelleher, N. L. *Anal. Chem.* **2004**, *76*, 196A–203A.
9. Bogdanov, B.; Smith, R. D. *Mass Spectrom. Rev.* **2005**, *24*, 168–200.
10. Collier, T. S.; Hawkridge, A. M.; Georgianna, D. R.; Payne, G. A. A.; Muddiman, D. C. *Anal. Chem.* **2008**, *80*, 4994–5001.
11. Boyne, M. T.; Garcia, B. A.; Li, M. X.; Zamdborg, L.; Wenger, C. D.; Babai, S.; Kelleher, N. L. *J. Proteome Res.* **2009**, *8*, 374–379.
12. Millea, K. M.; Krull, I. A.; Cohen, S, A.; Gebler, J. C.; Berger, S. J. *J. Proteome Res.* **2006**, *5*, 135–146.
13. Lee, J. E.; Kellie, J. F.; Tran, J. C.; Tipton, J. D.; Catherman, A. D.; Thomas, H. M; Ahlf, D. R.; Durbin, K. R.; Vellaicamy, A.; Ntai, I.; Marshall, A. G.; Kelleher, N. L. *J. Am. Soc. Mass Spectrom.* **2009**, *20*, 2183–2191.
14. Wynne, C.; Fenselau, C.; Demirev, P. A.; Edwards, N. J. *Anal. Chem.* **2009**, *81*, 9633–9642.
15. Demirev, P. A.; Feldman, A. B.; Kowalski, P.; Lin, J. S. *Anal. Chem.* **2005**, *77*, 7455–7461.

16. Fagerquist, C. K.; Garbus, B. R.; Miller, W. G.; Williams, K. E.; Yee, E.; Bates, A. H.; Boyle, S.; Harden, L. A.; Cooley, M. B.; Mandrell, R. E. *Anal. Chem.* **2009**, *7*, 2717–2725.
17. Macek, B.; Waanders, L. F.; Olsen, J. V.; Mann, M. *Mol. Cell Proteomics* **2006**, *5*, 949–958.
18. Wynne, C.; Edwards, N. J.; Fenselau, C. *Proteomics* **2010**, *10*, in press.
19. Edwards, N. J.; Pineda, R. Rapid Microorganism Identification Database (RMIDb). *54th Annual Conference of the American Society for Mass Spectrometry*, Seattle, WA May 28−June 1, 2006.
20. Altschul, S. F.; Gish, W.; Miller, W.; Myers, E. W.; Lipman, D. J. *J. Mol. Biol.* **1990**, *215*, 403–410.
21. Dereeper, A.; Guignon, V.; Blanc, G.; Audic, S.; Buffet, S.; Chevenet, F.; Dufayard, J.-F.; Guindon, S.; Lefort, V.; Lescot, M.; Claverie, J.-M.; Gascuel, O. *Nucleic Acids Res.* **2008**, *36*, W465–W469.
22. Perry, R. H.; Cooks, R. G.; Noll, R. J. *Mass Spectrom Rev.* **2008**, *27*, 661–699.
23. Yergey, A.; Heller, D.; Hansen, G.; Cotter, R.; Fenselau, C. *Anal. Chem.* **1982**, *55*, 353–356.
24. Zabrouskov, V.; Senko, M. W.; Du, Y.; Leduc, R. D.; Kelleher, N. L. *J. Am. Soc. Mass Spectrom.* **2005**, *16*, 2027–2038.
25. Liu, X.; Inbar, Y.; Dorrestein, P.; Wynne, C.; Edwards, N.; Souda, P.; Whitelegge, J. P.; Bafna, V.; Pevzner, P. A. *Mol. Cell. Proteomics* **2010**, in press.
26. Kessner, D.; Chambers, M.; Burke, R.; Agus, D.; Mallick, P. *Bioinformatics* **2008**, *24*, 2534–2536.
27. Delcher, A. L.; Bratke, K. A.; Powers, E. C.; Salzberg, S. L. *Bioinformatics* **2007**, *23*, 673–679.
28. Delcher, A. L. Personal communication, 2010.
29. Tobias, J. W.; Shrader, T. E.; Rocap, G.; Varshavsky, A. *Science* **1991**, *254*, 1374–1377.
30. Meng, F.; Cargile, B. J.; Miller, L. M.; Forbes, A. J.; Johnson, J. R.; Kelleher, N. L. *Nat. Biotechnol.* **2001**, *19*, 952–957.
31. Scherl, A.; Shaffer, S. A. A.; Taylor, G. K. K.; Hernandez, P.; Appel, R. D. D.; Binz, P.-A. A.; Goodlett, D. R. R. *J. Am. Soc. Mass Spectrom.* **2008**, *19*, 891–901.
32. Mount, D. W. *Bioinformatics: Sequence and Genome Analysis*; Cold Spring Harbor Laboratory Press: 2004.
33. Frank, A. M.; Pesavento, J. J.; Mizzen, C. A.; Kelleher, N. L.; Pevzner, P. A. *Anal. Chem.* **2008**, *80*, 2499–2505.
34. Montecchi-Palazzi, L.; Beavis, R.; Binz, P.-A.; Chalkley, R. J.; Cottrell, J.; Creasy, D.; Shofstahl, J.; Seymour, S. L.; Garavelli, J. S. *Nat. Biotechnol.* **2008**, *26*, 864–866.
35. Garavelli, J. S. Personal communication, 2010.

Chapter 9

Matrix Assisted Laser Desorption Ionization Ion Mobility Time-of-Flight Mass Spectrometry of Bacteria

Juaneka M. Hayes,[a] Louis C. Anderson,[b] J. Albert Schultz,[c] Michael Ugarov,[c] Thomas F. Egan,[c] Ernest K. Lewis,[c] Virginia Womack,[c] Amina S. Woods,[d] Shelley N. Jackson,[d] Robert H. Hauge,[e] Carter Kittrell,[e] Steve Ripley,[e] and Kermit K. Murray*,[a]

[a]Department of Chemistry, Louisiana State University, Baton Rouge, LA 70803
[b]U. S. Army, Dugway Proving Ground, Salt Lake City, UT 84022
[c]Ionwerks Inc., Houston, Texas 77030
[d]National Institute on Drug Abuse (NIDA) Intramural Research Program (IRP), National Institutes of Health (NIH), Baltimore, MD 21224
[e]Department of Chemistry, Rice University, Houston, Texas 77005
***kkmurray@lsu.edu**

Ion mobility mass spectrometry (IMMS) was combined with matrix assisted laser desorption ionization (MALDI) for the analysis of whole cell bacteria. Whole cell *Bacillus subtilis* ATCC 6633 and *Escherichia coli* strain W ATCC 9637 bacteria were prepared with a 1:2 analyte to matrix (CHCA) ratio and deposited using the dried-droplet method. Matrix-assisted laser desorption ionization time-of-flight mass spectrometry (MALDI-TOF MS) studies and matrix assisted laser desorption ionization ion mobility time-of-flight mass spectrometry (MALDI-IM-TOF MS) were conducted in parallel to assess the effectiveness of MALDI-IM-TOF MS for microorganism identification. Ribosomal proteins from *Escherichia coli* strain W ATCC 9637 were observed and assigned using the Rapid Microorganism Identification Database. Isoforms of lipopeptide products and a lantibiotic from the *Bacillus subtilis* ATCC 6633 species were found in the range 1000–3500 *m/z* and were identified using MALDI-IM-TOF MS and compared to MALDI-TOF MS data for confirmation. Vacuum ultraviolet (VUV) post-ionization MALDI-IM-TOF MS showed that additional information on lipopeptides could be obtained and used for identification.

Introduction

Mass spectrometry was first used to analyze bacteria in 1975 (*1*); results obtained from the investigation identified the peaks as phospholipids and ubiquinone products. Since then, advances in soft-ionization techniques, particularly matrix-assisted laser desorption ionization (MALDI), have led to broader use of mass spectrometry for the study of bacteria (*2*). MALDI typically produces singly charged ions, increasing sensitivity and reducing spectral congestion, resulting in rapid analysis. The characterization of bacteria using matrix-assisted laser desorption ionization time-of-flight mass spectrometry (MALDI-TOF MS) has been demonstrated by a number of groups (*1*, *3–8*). The MALDI approach to bacteria detection is attractive because of its low detection limit and tolerance for impurities. Identification is based upon characteristic peaks, termed biomarkers, that are attributed to molecules such as proteins, peptides, DNA, lipids, polysaccharides, and lipopolysaccharids (*9*). Spectra can be used to generate a reference library in which pattern matching or algorithms can be applied to compare spectra in order to determine identification. Most identifications are based on highly abundant and conserved ribosomal proteins which tend to dominate bacterial mass spectra (*10*). Using these approaches, MALDI has demonstrated the ability to identify bacteria at the species and genus levels and strains within a species (*10–12*).

It is standard practice to identify microorganisms using in-house databases or by using public database searches (*3*, *13*). There are some molecules found in bacteria that are conserved among different species, which makes distinguishing one species from another difficult when investigating complex mixtures (*5*, *14*, *15*). Analyzing complex mixtures can also be difficult due to spectral congestion and interferences due to isobaric lipids, peptides and other biomolecules. In addition, if bacteria are to be identified from collected material such as bioaerosols, significant background interferences may be encountered. To address these issues, there is a need for rapid analysis methods that have the ability to differentiate one bacterial species from one another based upon their characteristic profiles even while the biomarkers are obscured in mixtures of bacteria or background interferences.

Ion mobility spectrometry (IMS) is a fast gas phase technique that separates ions based upon their charge and collision cross section (*16*). The sample to be investigated is ionized and the ions enter the mobility drift tube. In the drift tube, ions are exposed to a weak electric field and a buffer gas at a pressure of a few Torr. Larger ions with greater collision cross sections interact more with the buffer gas than the smaller ions, thus smaller ions traverse the drift faster than larger ions. This means that ions of the same molecular weight but different size can be separated.

Ion mobility spectrometry combined with mass spectrometry (IMMS) is a useful tool for investigating the kinds of complex mixtures that can be encountered in bacterial biomarker identification, both background interferences and complex mixtures of bacteria (*17*). Both electrospray ionization (ESI) and MALDI have been used for analyte introduction for IMMS investigations. But for rapid bacteria identification, MALDI is more practical because of faster sample preparation, its tolerance to impurities, and its ability to produce singly charged ions, which can reduce spectral congestion. In IMMS, ions drift in the presence of a weak electric field through an intermediate pressure buffer gas (typically helium) and are separated according to the ratio of their size to charge, as in IMS. Ions with a low size to charge ratio encounter fewer collisions with the buffer gas and traverse the cell more quickly. This separation occurs on a millisecond time scale and is known as an ion mobility drift time (*16*). Different classes of molecules, which are common in biological samples and complex mixtures, such as peptides, lipids, carbohydrates, and nucleic acids are separated in the first dimension by size-to-charge and then in the second dimension by mass-to-charge.

The plot of ion mobility drift time as a function of *m/z* is a mass-mobility correlation represented in two-dimensional space for multiple components as groupings of peaks with different slopes that are called trend lines. Ions of a particular biomolecular class tend to lie on a single trend line. Each biomolecular class has different collision cross sections, oligonucleotides > carbohydrates > peptides > lipids (*18*, *19*), so when investigating a complex mixture results show multiple trend lines (*20*, *21*), thus, allowing initial identification of molecular classes (*18*). By using IMMS, identification of multiple biomolecular ions can be achieved at once, which decreases identification times.

IM cells can be interfaced with a number of different mass analyzers: quadrupole, ion trap, fourier transform ion cyclotron resonance (FTICR), and time-of-flight (*17*). The quadrupole is a widely used analyzer in MS because of its simplicity and low cost. This analyzer is excellent for portable MS but has a limited mass resolution, small *m/z* range, and limited scan speed. Due to its scan speed, a quadrupole is more suited for selective isomeric and isobaric separations in which the mass spectrometer is operating in the single ion monitoring mode (*16*). An ion-trap is sensitive and simple to use. However, ions must be accumulated before entering the ion-trap and it has a limited scan speed, low dynamic range, small *m/z* range, and limited mass resolution. The FTICR mass analyzer has excellent mass resolution and mass accuracy although it is not suitable for portable or transportable instruments and the scan speed is relatively slow (*16*). Time-of-flight (TOF) mass spectrometers have a large mass range, high transmission efficiency, high acquisition rate, good resolution and accuracy. TOFs are also simple, inexpensive, and can achieve a high scan rate. Mass spectra can be acquired in microseconds. Using IMS with TOF allows thousands of mass spectra to be obtained for each ion mobility spectrum (*16*).

IMS has been used as a tool for bacterial biomarker profiling (*22–24*). Snyder and co-workers used IMS to indirectly detect *E. coli* (ATCC 11303) by monitoring the reaction of in vivo *E. coli* β-galactosidase enzyme with (o-nitro-phenyl) β-D-galactopyranoside (ONPG), an enzyme assay used to detect water contaminated with *Enterobacteriaceae* (*23*). Similarly, IMS with thermal desorption for sample introduction was used to generate fingerprints and differentiate between bacterial strains of whole cell bacteria directly from colonies on agar plates (*22*). Using IMS in conjunction with MS, more than 200 metabolites from *E. coli* cultures were detected (*25*).

In this work, we investigated MALDI-IM-TOF MS and MALDI-TOF MS. Initially we tested IMMS for bacteria identification and then used this technique to study common bacteria. *Escherichia coli* and *Bacillus subtilis* are both well studied using mass spectrometry and have been investigated by a number of groups (*26–30*).

Experimental

Sample Preparation

Lyophilized bacterial species were suspended in an organic and acidic mixture and vortexed before being deposited onto the target. Matrix was added and allowed to co-crystallize with the bacterial species which was then analyzed using MALDI-TOF MS and MALDI-IM-TOF MS. Peaks from the MALDI-TOF MS and MALDI-IM-TOF MS spectra were searched in the Rapid Microorganism Identification Database for bacterial fingerprinting.

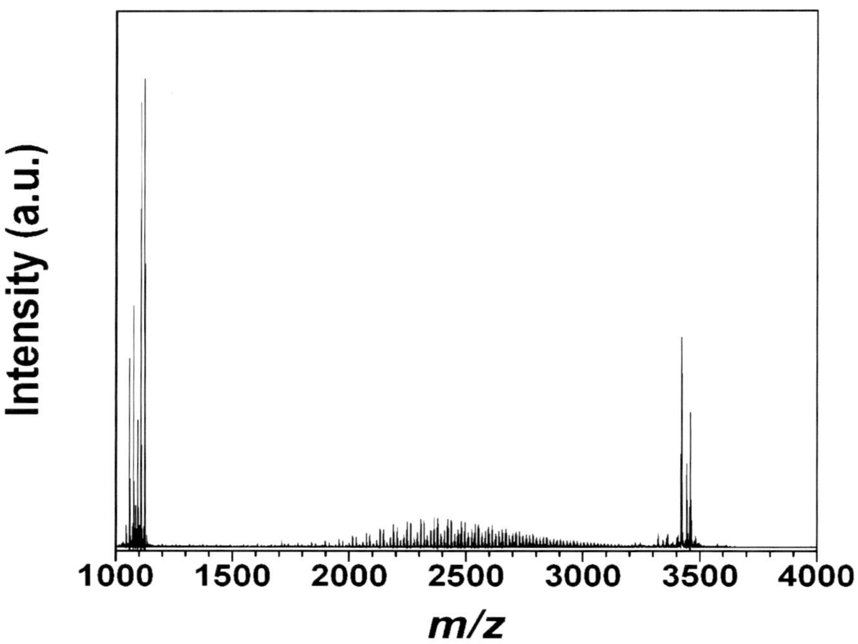

Figure 1a. MALDI mass spectra of B. subtilis ATCC 6633 using α-cyano-4-hydroxycinnamic acid (CHCA) matrix.

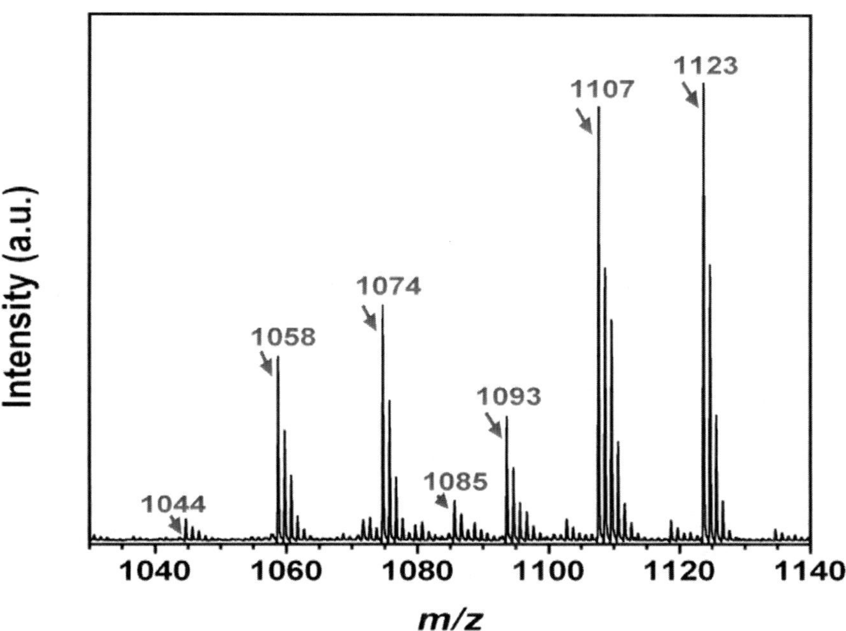

Figure 1b. Expanded view of B. subtilis ATCC 6633 lipopeptide products. (see color insert)

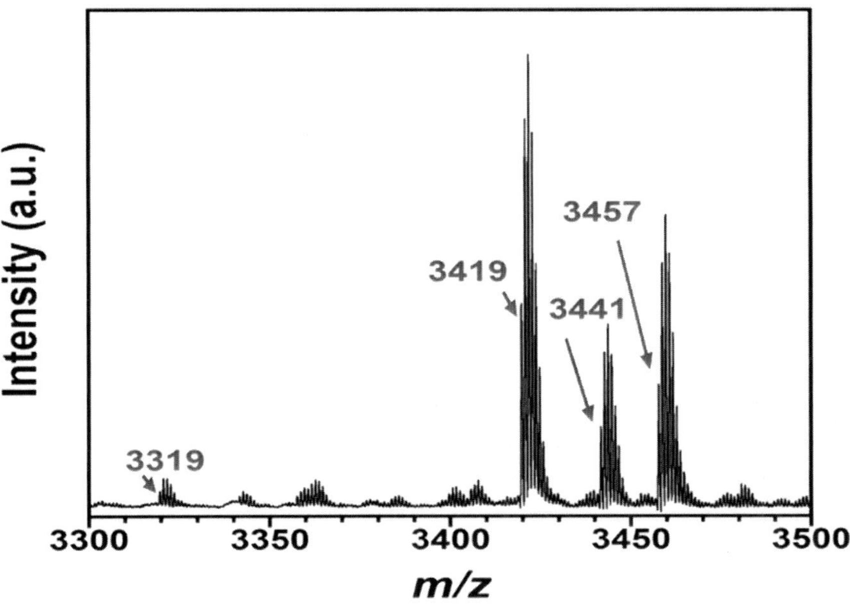

Figure 1c. Expanded view of B. subtilis ATCC 6633 subtilin with adducts and a protecting group. (see color insert)

Ion Mobility Mass Spectrometer

The matrix-assisted laser desorption ionization ion mobility time-of-flight mass spectrometer (Ionwerks, Houston, TX) has been described in detail previously (*18*). Briefly, the system is equipped with a diode pumped Q-switched, 349 nm (third harmonic) Nd:YLF laser (Crystalaser, Reno, NV) used at a repetition rate of 200 Hz (maximum repetition rate of 1 kHz). The beam was focused using a 15 cm focal length quartz lens that produced a spot size of approximately 60 µm. The sample is at ground, which allows the high voltage bias to pull the ions through the He gas. The mass spectrometer is floated at this bias voltage. Ions desorbed by the laser drift for 15 cm under a constant electric field in the mobility cell that was maintained at a pressure of 3 Torr helium. After each laser pulse, the ions drift to the end of the mobility cell, which is biased by 1900 V applied to a resistive divider network connected between the sample plate and the exit of the mobility spectrometer. At the end of the mobility cell, ions pass through a 0.5 mm orifice into a differentially pumped region before being orthogonally accelerated into the 40 cm reflectron TOF mass spectrometer. A mobility resolution of 30 (FWHM) and a mass resolution of 3000 (FWHM) at 1000 *m/z* was obtained. Ions were detected using a microchannel plate and four-anode detector. A time-to-digital-converter in ion-counting mode was used to acquire the signal. Data was plotted as two-dimensional contour plots of signal as a function of mobility and *m/z* using IDL software (Research Systems, Boulder, CO).

Vacuum ultraviolet (VUV) post-ionization was performed with a 157 nm fluorine excimer laser. The 349 nm desorption laser beam was focused onto the target and the VUV laser beam was parallel to the target several millimeters away. The repetition rate of the lasers was set to 200 Hz and the delay time between the two lasers is 500 µs. The VUV laser beam was focused to a spot size of 0.5 mm x 1 mm, using a custom-built enclosure with adjustable controls purged with nitrogen. This configuration allows the VUV laser to intercept and post ionize the neutrals in the laser desorbed plume. The ions were separated in the ion mobility cell and then by the TOF, as described above.

MALDI TOF Mass Spectrometry

MALDI-TOF MS analysis was carried out using an UltrafleXtreme MALDI-TOF/TOF mass spectrometer (Bruker Daltonics, Billerica, MA, USA) with a 1 kHz Smartbeam II laser with a broad band resolving power of 40,000 and 1 ppm accuracy.

Reagents

Lyophilized *Bacillus subtilis* ATCC 6633, *Escherichia coli* strain W ATCC 9637, and α-cyano-4-hydroxycinnamic acid (CHCA) were purchased from Sigma Aldrich (St. Louis, MO) and used without further purification. HPLC grade acetonitrile (ACN 99.9%) and trifluoroacetic acid (TFA; 100%) were purchased from Fisher Scientific (Pittsburgh, PA). A total of 20 mg/ml of the cells were

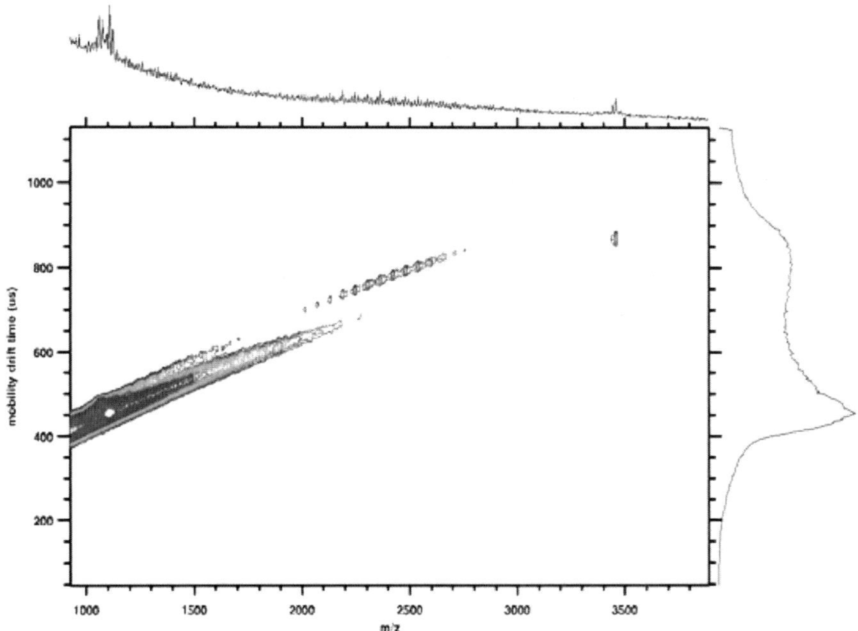

Figure 2. UV MALDI-IM-TOF MS 2-D contour plot of whole cell B. subtilis ATCC 6633. (see color insert)

suspended in 1:1 ACN/0.1% TFA. Saturated matrix solutions were prepared by dissolving 30 mg of CHCA matrix in 1 mL of 1:1 ACN/0.1% TFA. The saturated matrix was mixed with the suspension in a 1:2 ratio of bacterial suspension matrix solution and a 1 μl volume was deposited on the target and allowed to dry. Samples were prepared in the same manner for both MALDI-TOF MS and MALDI-IM-TOF MS experiments.

Results

MALDI-TOF MS and MALDI-IM-TOF MS experiments were performed on dried-droplet intact whole cell *Bacillus subtilis* ATCC 6633 and *Escherichia coli* strain W ATCC 9637. The information obtained from the MALDI spectra is compared with spectra obtained from the MALDI-IM-TOF MS experiments. Peaks from the *Escherichia coli* strain W ATCC 9637 spectra were searched for possible identification using Rapid Microorganism Identification Database. Additionally, VUV post-ionization MALDI-IM-TOF MS experiments were performed to obtain more information from the intact bacteria.

B. subtilis ATCC 6633

A MALDI-TOF mass spectrum of *Bacillus subtilis* ATCC 6633 is shown in Figure 1a. As seen in Figure 1a, a number of peaks in the range of 1–4 kDa were

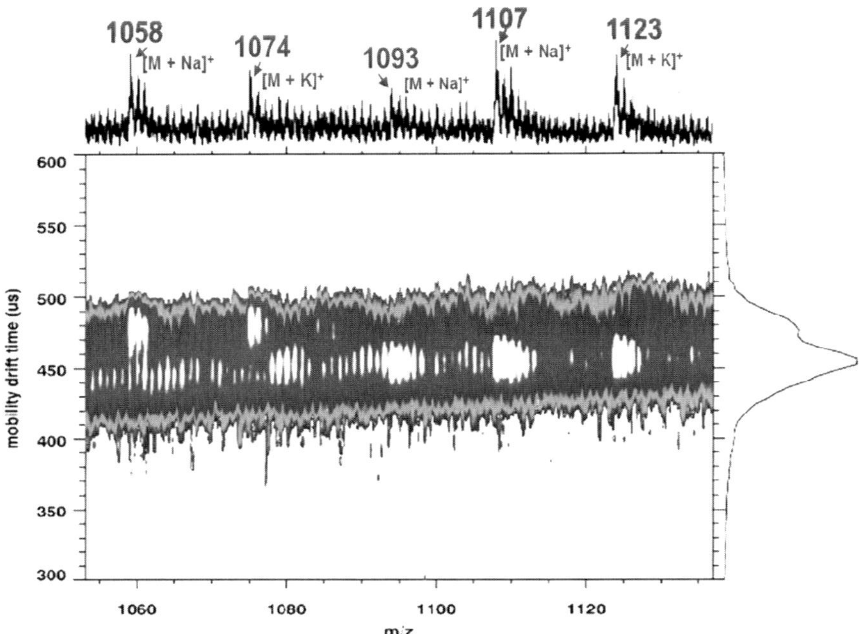

Figure 3. UV MALDI-IM-TOF MS 2-D contour plot of lipopeptide products with Na^+ and K^+ adducts from whole cell B. subtilis ATCC 6633. (see color insert)

observed for *Bacillus subtilis* ATCC 6633. The first cluster of peaks near 1100 *m/z* in Figure 1a is attributed to two classes of isoforms of lipopeptides known as surfactins and mycosubtilins (iturin family) which have been previously identified using MALDI MS (*26*, *27*, *31–33*). Lipopetides are a class of non-ribosomally generated amphiphilic peptides that are categorized according to their structure and activity and can be found in a number of different species of *Bacillus*. Surfactins are composed of a β-hydroxy fatty acid (C_{13} to C_{16}) linked to a cyclic lipoheptapeptide with n = 9-11, where n is the number of CH_2 groups (*26*). Mycosubtilins are very similar to surfactins but contains a β-amino fatty acid sequence (C_{14} to C_{17}), n = 11-13. The peak detected at 1044 is surfactin b-C_{15} with a sodium adduct and the ions at 1058 and 1074, surfactin a-C_{15} with sodium and potassium adducts (Figure 1b). The peaks at 1085, 1107, and 1123 *m/z* (Figure 1b) are representative of H^+, Na^+, and K^+ mycosubtilin-C_{17}, respectively. The peak at 1093 corresponds to mycosubtilin-C_{16} with a Na^+ adduct (Figure 1b). The broad distribution of peaks from 2000 *m/z* to near 3000 *m/z* in Figure 1a results from the ionization of the peptidoglycan layer of the bacterial cell wall separated by 14 Da (CH_2 group). The third cluster of peaks near 3500 *m/z* shown in Figure 1c correspond to a 32-amino-acid pentacyclic lantibiotic ribosomally synthesized as a prepeptide which undergoes posttranslational modification resulting in the mature protein subtilin, which is a lantibiotic that can be found in different strains of *B. subtilis* (*26*, *27*, *32*). The signal at 3319 *m/z* is the protonated ribosomally synthesized lantibiotic subtilin and the most intense peak seen at 3419 *m/z* is attributed to lantibiotic subtilin with a N-terminally succinylated subtilin. Na^+

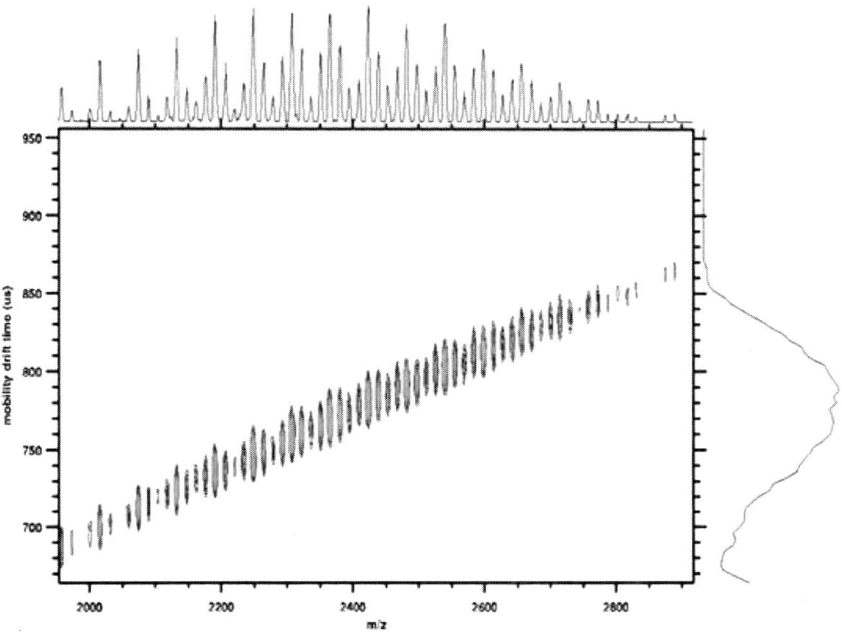

Figure 4. UV MALDI-IM-TOF MS 2-D contour plot of fatty acids separated by 14 Da from whole cell B. subtilis ATCC 6633. (see color insert)

and K^+ adducts of the succinylated subtilin are observed at 3441 *m/z* and 3457 *m/z*, respectively.

The two-dimensional MALDI ion mobility mass spectrometry contour plot of *Bacillus subtilis* ATCC 6633 is shown in Figure 2. The mass spectrum was obtained using a 349 nm laser at a repetition rate of 200 Hz. The plot above the top x-axis is the mass spectrum obtained by integrating the signal over all mobility times and the plot on the right hand y-axis is the ion mobility trace obtained by integrating over the *m/z* values. The plot shows three groupings of peaks in the range between 1000 and 3500 *m/z* on two distinct trend lines. From previous work (*18*, *34*), it can be inferred that the upper trend line corresponds to lipids whereas the lower trend line corresponds to peptides. The strong signal on the lower trend line 1100 *m/z* results from lipopeptides. In the middle of the contour plot, the features on the upper trend line correspond to lipids. The island of signal at the upper right of the plot corresponds to subtilin.

The first cluster of peaks in the mass spectrum of the 2-D contour plot of Figure 2 is shown in an expanded view in Figure 3. The pattern of the first cluster of peaks in the mass spectrum of the 2-D contour plot is typically found for lipopeptides, representative of surfactin and mycosubtilin of the iturin family (*26*). One can see that the two most intense peaks at 1058 *m/z* and 1074 *m/z* with a mobility drift time of 470 μs in the first cluster are attributed to surfactin-C_{15} with sodium and potassium adducts just as in the MALDI-TOF MS spectrum shown in Figure 1b. Isoforms mycosubtilin-C_{16} and mycosubtilin-C_{17} are observed at 1093 and 1107 *m/z*, respectively, both with Na^+ adducts and a mobility drift time of 440 μs. The

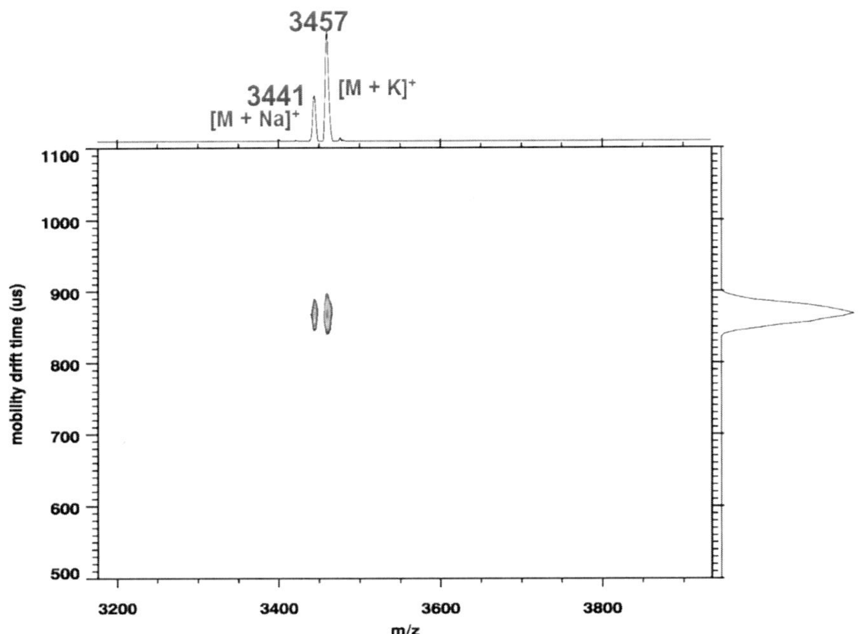

Figure 5. UV MALDI-IM-TOF MS 2-D contour plot of whole cell B. subtilis ATCC 6633. The peaks correspond to N-terminal succinylated subtilin with Na^+ and K^+ adducts. (see color insert)

peak at 1123 *m/z* with a mobility drift time of 440 μs is assigned as the potassium adduct of mycosubtilin-C_{17}, as observed in Figure 1b.

The second cluster of peaks in the middle of the spectrum from Figure 2, from approximately 1900 *m/z* to near 3000 *m/z* with a mobility drift time range of 670 μs to 790 μs, is shown in an expanded view in Figure 4. These peaks are separated by CH_2, 14 Da, and are attributed to the peptidoglycan layer of the bacterial cell wall.

In Figure 5, the expanded view of the third cluster near 3450 *m/z* with a mobility drift time of 870 μs from Figure 2 is shown. The peak seen at 3441 *m/z* is assigned to the $[M + Na]^+$ peak of succinated subtilin; the K^+ adduct appears at 3457 *m/z*. The subtilin or succinated subtilin molecular ion peaks were not observed as in Figure 1c. The mass spectra for subtilin as well as the lipopeptides are in agreement with spectra shown in current mass spectrometry literature (*26*).

To obtain additional information, VUV post-ionization MALDI-IM-TOF MS was also used for the analysis of *B. subtilis* ATCC 6633. As expected, the groupings of the isoforms of surfactins and mycosubtilins observed using UV MALDI-IM-TOF MS have the same mobility and lie on the same trend lines, although a few additional lipopeptides were observed, Figure 6. These additional lipopeptides are associated with peaks in the x-axis of the contour plot at 1079, 1090, and 1329 *m/z*. The ion peaks at 1079 and 1090 *m/z* lie on the same trend line with the aforementioned surfactins and mycosubtilins, suggesting they are isoforms of surfactin and mycosubtilin with a Na^+ and K^+ adduct,

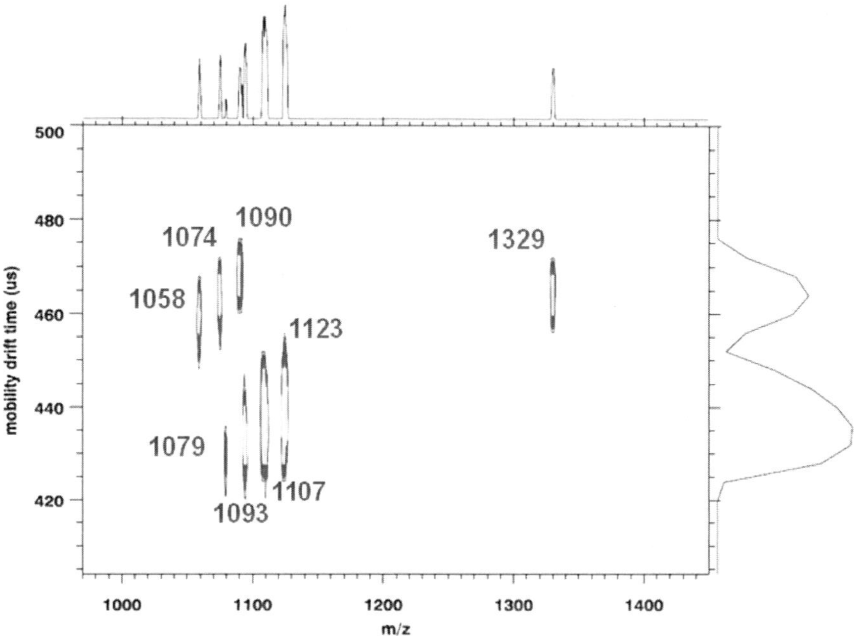

Figure 6. VUV post-ionization MALDI-IM-TOF MS 2-D contour plot of whole cell B. subtilis ATCC 6633. Additional peaks at 1079, 1090, and 1329 m/z were observed. (see color insert)

respectively. Although, the ion peak at 1329 *m/z* peak lies on the same trend line as the mycosubtilin family, it is yet to be identified. In the VUV post-ionization MALDI-IM-TOF MS spectrum of Figure 6, mycosubtilin-C_{17} with the K^+ adduct is observed at 1123 *m/z*.

E. coli W ATCC 9637

In Figure 7, the MALDI spectrum resulting from the analysis of *E. coli* strain W ATCC 9637 in CHCA is shown. This spectrum is similar to previous published reports of *E. coli* (*28*, *29*, *35*). The observed peaks are listed in Table 1. All *m/z* values between 4000 and 15000 m/z were searched against the Rapid Microorganism Identification Database (RMIDb) (*36*). The model (type of proteins) used for this search was Bacterial Ribosomal Proteins from all sources, which included Genbank, TrEMBL, SwissProt, RefSeq, Ventor Institute's Comprehensive Microbial Resource (CMR), and Glimmer3. The search was conducted using *m/z* values observed with a + 1 charge (assuming singly protonated). The error window selected for the search performed was ± 10. Of the observed peaks, most respond to ribosomal subunit proteins and four (5097, 5384, 6416, and 7280 *m/z*) have been previously identified as singly charge ribosomal proteins (*37*), while a few are Glimmer3 predictions. Glimmer3 is an algorithm that predicts protein sequences on bacterial genomes. These possible protein sequences are from a number of protein database sources. If Glimmer3 is

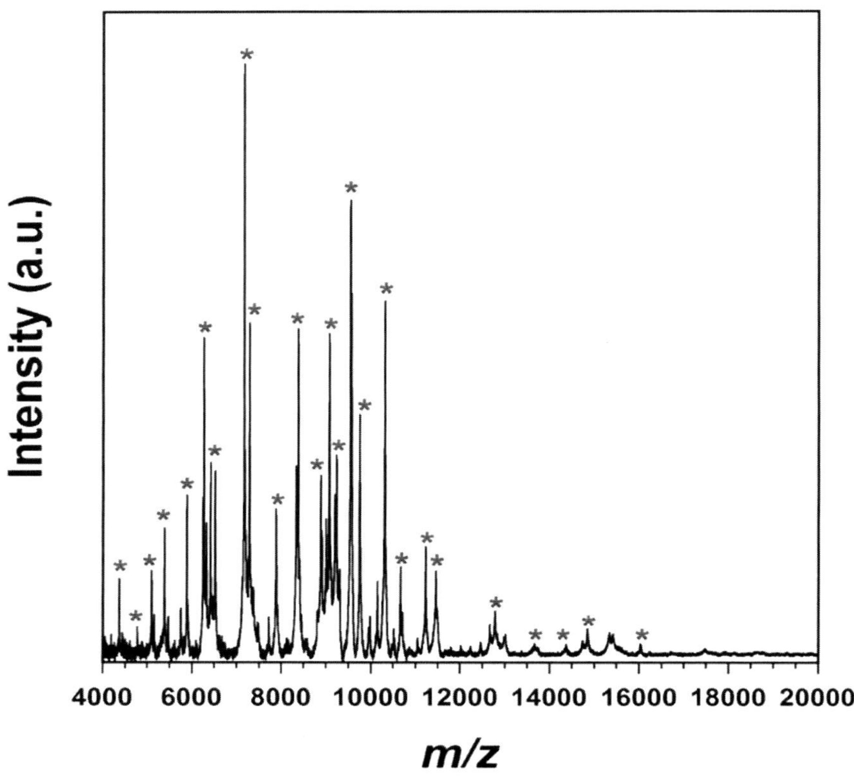

Figure 7. MALDI- TOF MS mass spectra of E. coli ATCC 9637 using α-cyano-4-hydroxycinnamic acid (CHCA). (see color insert)

the only matching protein sequence, this means there is no protein sequence that matches the given mass. There is a possibility that the protein sequence does not exist or it could be a real protein that is simply missing from the protein sequence database.

The 2-D contour plot resulting from the analysis of *E. coli* strain W ATCC 9637 is shown in Figure 8. The plot shows two trend lines on which corresponds to a total of 18 peaks. The peaks observed in the mass range of 1000 to 12000 *m/z* with a mobility time of 670 to 1780 μs are shown in Table 2. For this analysis, peaks of 4000 m/z or greater were searched against the Rapid Microorganism Identification Database using the model Bacterial Ribosomal Proteins from all sources. The search was conducted using *m/z* values observed with a +1 charge (assuming singly protonated). The error window chosen for the search performed was ± 10 Da. The RMIDb search of the peaks listed in Table 2 from the MALDI-IM TOF MS data yielded a Glimmer3 prediction at 4742 *m/z*. Most of the peaks from the MALDI-TOF MS data listed in Table 1 were identified as ribosomal proteins, while a few values were assigned as Glimmer 3 predicted sequences.

Table 1. Peaks observed from MALDI-TOF MS spectrum of *E. coli* ATCC 9637

Observed Mass	*Theoretical Mass*	*Protein Description*	*Organism*	*Accession number*
4366	4365	50 S Ribosomal protein L36	*E. coli* E24377A	A7ZSI8
5097	5096	30 S Ribosomal protein S22	*E. coli* E2348168	Z15486706
5384	5381	50 S Ribosomal protein L34	*E. coli* E2348168	A7ZTQ9
6416	6411	50 S Ribosomal protein L30	*E. coli* E24377A	A7ZSJ1
7164	7158	50 S Ribosomal protein L35	*E. coli* S88	B7MAS7
7280	7274	50 S Ribosomal protein L29	*E. coli* E24377A	A7ZSK1
7877	7872	50 S Ribosomal protein L31	*E. coli* E24377A	A7ZUF1
8376	8369	30 S Ribosomal protein S21	*E. coli* UT189	215488396
8884	8876	50 S Ribosomal protein L28	*E. coli* E24377A	A7ZT18
9544	9548	Glimmer3 prediction	*E. coli* 536	GL82531
10146	10138	30 S Ribosomal protein S15	*E. coli* E2438169	215488483
10309	10300	30 S Ribosomal protein S19	*E. coli* E2438169	215488616
11233	11229	Glimmer3 prediction	*E. coli* 536	GL8253
11460	11464	30 S Ribosomal protein S14	*E. coli* E2438169	218555864
12780	12770	50 S Ribosomal protein L18	*E. coli* E24377A	A7ZSJ3
14372	14365	50 S Ribosomal protein L29	*E. coli* E24377A	A7ZSK1

Table 2. Peaks observed from MALDI-IM-TOF MS spectrum of *E. coli* ATCC 9637

Observed Mass	*Theoretical Mass*	*Protein Description*	*Organism*	*Accession number*
1736				
1980				
2583				
3233				
3271				
3272				
3618				
3620				
4590				
4742	4750	Glimmer3 prediction	*E. coli* ATCC 8739	GL10468.1400
5065				
5070				
5084				
6372				
7230				
9133				
9168				
9474				

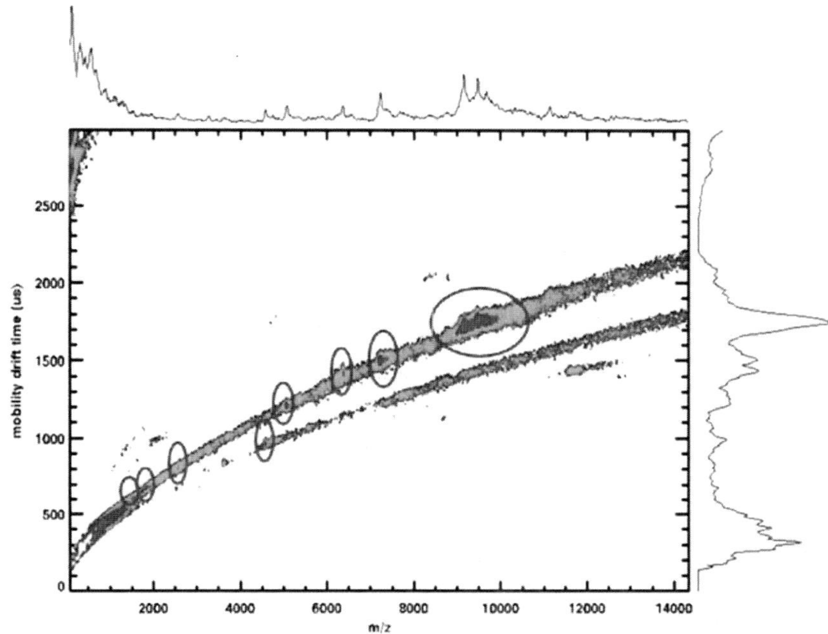

Figure 8. UV MALDI-IM-TOF MS 2-D contour plot of whole cell E. coli ATCC 9637. (see color insert)

Discussion

Using MALDI IMMS, it was possible to separate in a *Bacillus subtilis* 6633 isoforms of a class of non-ribosomally generated lipopetides (surfactins, mycosubtilins), a ribosomally synthesized lantibiotic (subtilin), and a polymer-like pattern thought to be associated with the peptidoglycan layer of the cell wall. Furthermore, an additional surfactin isoform (1090 *m/z*) and an additional mycosubtilin isoform at (1079 *m/z*) were observed when using VUV post-ionization MALDI-IM-TOF MS, as well as a MH^+, molecular ion, at 1329 *m/z* that most likely belongs to the mycosubtilin family, suggested by its position on the trend line with other mycosubtilins. The ion at 1329 *m/z* was only observed in the VUV post-ionization MALDI-IM TOF MS experiments. MALDI-IM-TOF MS was also effective in separating proteins from *E. coli* W 9637.

The strategy of MALDI-IM-TOF MS studies in parallel with the MALDI-TOF MS studies highlights the added advantage of the ion mobility dimension. In Figure 2, the two different trend lines observed in the two-dimensional fingerprint of the *B. subtilis* ATCC 6663 establishes the presence of different molecular classes of ions, which in this case are peptides and lipids, before any type of database inquiry. In Figure 3, the divergence of ions of the same biomolecular class from the trend line is the result of structural differences (*20*). In this case, both lipopeptides, the surfactins and mycosubtilins, have similar drift times but the mycosubtilins lie slightly below the lipopeptide trend line due to its more compact structure.

The surfactins are linked by a β-hydroxy fatty acid whereas mycosubtilins have a β-amino fatty acid linkage. The unidentified weak signals in the 1-D mass portion of the plot in the range 1050 to 1150 *m/z* are attributed to isoforms of surfactins and mycosubtilins, which have also been noted in recent work (*26*, *27*). The difference in the mobility drift time of the surfactin and mycosubtilins can be seen more distinctly in the VUV MALDI-IM-TOF MS data of Figure 6. Figure 6 also shows that three additional peaks were observed when using VUV MALDI-IM-TOF. One of the additional peaks observed is 1090 *m/z*. This lies on the same trend line as the previously identified 1058 and 1074 *m/z* peaks. The other two additional peaks observed at 1079 and 1329 *m/z* lie on the same trend line as the mycosubtilins. These additional peaks lie on the surfactin and mycosubtilin trend lines, suggesting they are isoforms of the surfactin and mycosubtilin families.

Conclusion

This study demonstrated that the bacteria *B. subtilis* ATCC 6633 and *E. coli* ATCC 9637 can be detected from lyophilized bacterial cells using MALDI-IM-TOF MS. Isobaric lipids, peptides, and proteins are separated on IMMS trend lines, which allows the type of biomolecule to be distinguished prior to *m/z* separation. The structural differences of the lipopeptides surfactin and mycosubtilin, of *B. subtilis* 6633 were detected because of the deviation of the mycosubtilin below the lipopeptide trend line. It was also possible to observe additional peaks using VUV MALDI-IM-TOF MS that were assigned as surfactin and mycosubtilin isoforms because of their presence on the surfactin and mycosubtilin trend lines. Proteins detected from *E. coli* W 9637 lie on the same trend line; however, it was not possible to identify them. The *m/z* values of the peaks observed were searched in the RMIDb. A Glimmer3 predicted sequence was obtained for one of the *m/z* values.

MALDI-IM-TOF MS can be used as a tool for biomarker identification that can provide additional information in the presence of different compound classes that complements MALDI- TOF and can provide additional insight when both tools are used together. The ability of MALDI-IM-TOF MS to separate isobaric species and improve signal to noise makes it ideal for routine identification. This technique can be used to analyze collected bioaerosols as has been demonstrated with MALDI-TOF MS (*38*) in the presence of background matrices and can be applied to bioaerosol mixtures.

Acknowledgments

This research is supported by U.S. Army Dugway Proving Ground under contract W911S6-07-P-0063.

References

1. Anhalt, J. P.; Fenselau, C. *Anal. Chem.* **1975**, *47*, 219–225.
2. Jackson, O. L., Jr. *Mass Spectrom. Rev.* **2001**, *20*, 172–194.

3. Demirev, P. A.; Ho, Y. P.; Ryzhov, V.; Fenselau, C. *Anal. Chem.* **1999**, *71*, 2732–2738.
4. Krishnamurthy, T.; Rajamani, U.; Ross, P.; Jabbour, R.; Nair, H.; Eng, J.; Yates, J.; Davis, M.; Stahl, D.; Lee, T. *J. Toxicol., Toxin Rev.* **2000**, *19*, 95.
5. Jarman, K. H.; Cebula, S. T.; Saenz, A. J.; Petersen, C. E.; Valentine, N. B.; Kingsley, M. T.; Wahl, K. L. *Anal. Chem.* **2000**, *72*, 1217–1223.
6. Conway, G. C.; Smole, S. C.; Sarracino, D. A.; Arbeit, R. D.; Leopold, P. E. *J. Mol. Microbiol. Biotechnol.* **2001**, *3*, 103–112.
7. Ryzhov, V.; Hathout, Y.; Fenselau, C. *Appl. Environ. Microbiol.* **2000**, *66*, 3828–3834.
8. Winkler, M. A.; Uher, J.; Cepa, S. *Anal. Chem.* **1999**, *71*, 3416–3419.
9. Krishnamurthy, T.; Ross, P. L.; Rajamani, U. *Rapid. Commun. Mass Spectrom.* **1996**, *10*, 883–888.
10. Demirev, P. A.; Fenselau, C. *Annu. Rev. Anal. Chem.* **2008**, *1*, 71–93.
11. Carbonnelle, E.; Mesquita, C.; Bille, E.; Day, N.; Dauphin, B.; Beretti, J.-L.; Ferroni, A.; Gutmann, L.; Nassif, X. *Clin. Biochem.* **2010**, in press (corrected proof).
12. Sauer, S.; Kliem, M. *Nat. Rev. Microbiol.* **2010**, *8*, 74–82.
13. Yao, Z.-P.; Demirev, P. A.; Fenselau, C. *Anal. Chem.* **2002**, *74*, 2529–2534.
14. Claydon, M. A.; Davey, S. N.; EdwardsJones, V.; Gordon, D. B. *Nat. Biotechnol.* **1996**, *14*, 1584–1586.
15. Wahl, K. L.; Wunschel, S. C.; Jarman, K. H.; Valentine, N. B.; Petersen, C. E.; Kingsley, M. T.; Zartolas, K. A.; Saenz, A. J. *Anal. Chem.* **2002**, *74*, 6191–6199.
16. Kanu, A. B.; Dwivedi, P.; Tam, M.; Matz, L.; Hill, H. H. *J. Mass Spectrom.* **2008**, *43*, 1–22.
17. Bohrer, B. C.; Merenbloom, S. I.; Koeniger, S. L.; Hilderbrand, A. E.; Clemmer, D. E. *Annu. Rev. Anal. Chem.* **2008**, *1*, 293–327.
18. Woods, A. S.; Schultz, J. A.; Ugarov, M.; Egan, T.; Koomen, J.; Gillig, K. J.; Fuhrer, K.; Gonin, M. *Anal. Chem.* **2004**, *76*, 2187–2195.
19. Koomen, J. M.; Routolo, B. T.; Gillig, K. J.; McClean, J. A.; Russell, D. H.; Kang, M.; Dunbar, K. R.; Fuhrer, K.; Gonin, M.; Schultz, J. A. *Anal. Bioanal. Chem.* **2002**, *373*, 612–617.
20. Russell, D. H.; Gillig, K. J.; Verbeck, G. F.; Ruotolo, B. T.; Sawyer, H. A. *J. Biomol. Tech.* **2002**, *13*, 56–61.
21. Henderson, S. C.; Valentine, S. J.; Counterman, A. E.; Clemmer, D. E. *Anal. Chem.* **1999**, *71*, 291–301.
22. Vinopal, R. T.; Jadamec, J. R.; deFur, P.; Demars, A. L.; Jakubielski, S.; Green, C.; Anderson, C. P.; Dugas, J. E.; DeBono, R. F. *Anal. Chim. Acta* **2002**, *457*, 83–95.
23. Snyder, A. P.; Shoff, D. B.; Eiceman, G. A.; Blyth, D. A.; Parsons, J. A. *Anal. Chem.* **1991**, *63*, 526–529.
24. Dwivedi, P.; Wu, P.; Klopsch, S. J.; Puzon, G. J.; Xun, L.; Hill, H. H. *Metabolomics* **2008**, *4*, 63–80.
25. Dwivedi, P.; Wu, P.; Klopsch, S.; Puzon, G.; Xun, L.; Hill, H. *Metabolomics* **2008**, *4*, 63–80.

26. Leenders, F.; Stein, T. H.; Kablitz, B.; Franke, P.; Vater, J. *Rapid. Commun. Mass Spectrom.* **1999**, *13*, 943–949.
27. Stein, T. *Rapid. Commun. Mass Spectrom.* **2008**, *22*, 1146–1152.
28. Wang, Z.; Russon, L.; Li, L.; Roser, D. C.; Long, S. R. *Rapid. Commun. Mass Spectrom.* **1998**, *12*, 456–464.
29. Jones, J. J.; Stump, M. J.; Fleming, R. C.; Lay, J. O.; Wilkins, C. L. *Anal. Chem.* **2003**, *75*, 1340–1347.
30. Ryzhov, V.; Fenselau, C. *Anal. Chem.* **2001**, *73*, 746–750.
31. Stein, T.; Borchert, S.; Conrad, B.; Feesche, J.; Hofemeister, B.; Hofemeister, J.; Entian, K. *J. Bacteriol.* **2002**, *184*, 1703–1711.
32. Stein, T.; Entian, K. D. *Rapid. Commun. Mass Spectrom.* **2002**, *16*, 103–110.
33. Madonna, A. J.; Voorhees, K. J.; Taranenko, N. I.; Laiko, V. V.; Doroshenko, V. M. *Anal. Chem.* **2003**, *75*, 1628–1637.
34. Russell, D. H.; Gillig, K. J.; McLean, J. A.; Ruotolo, B. T. *Int. J. Mass Spectrom.* **2005**, *240*, 301–315.
35. Arnold, R. J.; Reilly, J. P. *Rapid. Commun. Mass Spectrom.* **1998**, *12*, 630–636.
36. Edwards, N. J. P. F. Presented at the 54th Conference of the Annual American Society of Mass Spectrometry, Seattle, WA, May 28−June 1, 2006.
37. Arnold, R. J.; Karty, J. A.; Ellington, A. D.; Reilly, J. P. *Anal. Chem.* **1999**, *71*, 1990–1996.
38. Murray, K. K.; Jackson, S. N.; Kim, J. *Rapid. Commun. Mass Spectrom.* **2005**, *2205*, 1725–1729.

Chapter 10

Single-Particle Aerosol Mass Spectrometry (SPAMS) for High-Throughput and Rapid Analysis of Biological Aerosols and Single Cells

Matthias Frank,* Eric E. Gard, Herbert J. Tobias,[1] Kristl L. Adams, Michael J. Bogan,[2] Keith R. Coffee, George R. Farquar, David P. Fergenson,[3] Sue I. Martin,[4] Maurice Pitesky,[5] Vincent J. Riot, Abneesh Srivastava,[6] Paul T. Steele, and Audrey M. Williams

Lawrence Livermore National Laboratory, 7000 East Ave., L-211, Livermore, CA 94551
***frank1@llnl.gov**

[1]Current address: Cornell University, Ithaca NY, 14853
[2]Current address: PULSE Institute, SLAC National Accelerator Laboratory, 2575 Sand Hill Road MS 59, Menlo Park CA 94025
[3]Current address: Livermore Instruments Inc., 6773 Sierra Court, Suite C, Dublin, CA 94568
[4]Current address: Simbol Mining Corp., 6920 Koll Center Parkway, Pleasanton, CA 94566
[5]Current address: University of California Davis, Davis, CA 95616
[6]Current address: Entegris, Inc., San Diego, CA 92131

At Lawrence Livermore National Laboratory, we have developed a single-particle aerosol mass spectrometry (SPAMS) system that can rapidly analyze individual micrometer-sized biological aerosol particles or cells that are sampled directly from air or a lab-generated aerosol into a mass spectrometer. As particles enter the SPAMS system, their aerodynamic size and fluorescence properties are measured before mass spectra from both positive and negative ions created by matrix-free laser desorption and ionization are recorded. All the correlated data obtained from a particle can be analyzed and classified in real-time. The SPAMS system is capable of discriminating, particle by particle, between bacterial spores, vegetative cells and other biological and non-biological background materials using the mass fingerprints obtained from those particles. In addition, selected species of bacteria can be discriminated from each other with this method. Here we describe the overall architecture of the SPAMS system and the related algorithms. We present selected results from applying the SPAMS technique to the analysis of biological agent simulants and single cells. We also describe results from first proof-of-concept experiments using SPAMS for the rapid screening of human effluents for tuberculosis. Lastly, we present results from a field study in a large airport using SPAMS to assess biological content in ambient aerosol.

Introduction

Biological warfare agents that can be released as aerosols containing pathogenic bacteria, bacterial spores, viruses or biological toxins are of major concern because they can be disseminated easily and quickly over wide areas and in potentially lethal doses. A host of detection strategies and a wide range of detection technologies have been developed over the last decades, but each has its limitations (*1*). The sensitive detection and identification of airborne biological particles in real-time (in seconds, ideally) is a critical capability required for an effective response to a bioterrorist attack. Sensitive, real-time detectors for biological aerosols would likely also be useful for certain medical screening or diagnosis applications in public health or point-of-care diagnostics.

Various forms of mass spectrometry have been explored for the detection and identification of microorganisms with promising results (for recent reviews see (*2*, *3*) and also other chapters in this book). Samples have included bacteria and viruses grown in the lab as well as biological aerosols collected from aerosol chambers or the environment during field tests involving intentional release of agent surrogates or field measurements of naturally occurring biological background.

One form of mass spectrometry that may be particularly suited to rapidly detect the presence of biological aerosols in ambient air is single-particle aerosol mass spectrometry. Aerosol mass spectrometry has been developed over the last decades by numerous groups under various names (*4–11*). In aerosol mass spectrometry, aerosols from ambient air or produced by aerosol generators are sampled straight from air into a mass spectrometer system and analyzed. Most aerosol mass spectrometers today analyze *individual*, micrometer or sub-micron sized particles. While the majority of groups utilizing aerosol mass spectrometry does so for environmental science applications, our group at LLNL and few other groups have explored and further developed aerosol mass spectrometry for biological detection.

Over the last decade, we have developed a high-throughout single-particle aerosol mass spectrometry (SPAMS) system that can rapidly analyze individual micrometer-sized biological aerosol particles or cells that are sampled directly from air or a lab-generated aerosol into a mass spectrometer. Originally, our system was called BAMS for BioAerosol Mass Spectrometer. As aerosol detection demonstrations were expanded to include other types of threats, such as explosives and chemicals, the system was renamed SPAMS to reflect the broader application range. One focus of this work was to increase the speed and sensitivity by several orders of magnitude compared to other types of aerosol mass spectrometry. The other focus was on developing a system that can operate in a totally autonomous fashion for extended periods of time while performing data analysis on board and in real time and continually surveying incoming single-particle data with dynamic alarm algorithms that adapt automatically to fluctuating aerosol backgrounds as measured by the system itself.

Compared to other aerosol mass spectrometers, SPAMS makes use of several newly developed components: a high efficiency particle inlet and aerodynamic lens combination to sample and focus aerosol particles, an advanced tracking and aerodynamic sizing region that predicts the location and speed of the particles inside the system and a two-band fluorescence pre-selection stage. While for earlier generations of SPAMS the final stage, a dual polarity (negative and positive) mass-spectrometer for single particle chemical analysis, was based on the mass spectrometer from a commercial instrument the latest SPAMS instrument is equipped with an LLNL-developed miniaturized mass spectrometer. In the SPAMS mass spectrometer particles are hit with a pulse from a UV desorption/ionization laser (triggered by the timing information from the tracking stage above) and mass spectra from both positive and negative ions created by matrix-free laser desorption and ionization are recorded. The dual-polarity mass spectrum recorded for each individual particle can be used as a signature.

While we originally developed SPAMS specifically for high-throughput and rapid analysis of biological aerosols for biodetection applications (*12*) several other applications were explored, subsequently. Those include using SPAMS for single-cell analysis (*13*) and biomedical applications (*14*, *15*), environmental applications (*16*), explosives detection (*17*), chemical agent detection (*18*), pesticide detection (*19*), and analysis of drugs (*20*, *21*). Finally, it was demonstrated that SPAMS can simultaneously detect different types of threat materials with the same system settings and pattern recognition library and has the potential to be a "universal threat detector" (*22*).

In this chapter, we first describe some of the SPAMS hardware and experimental details. We then discuss SPAMS results obtained from *Bacillus* spores and vegetative bacterial cells before describing automated single-particle data analysis algorithms based on pattern matching and rules trees as well as providing some details on alarm algorithms developed for SPAMS. The remainder of this chapter contains some results from proof-of-concept experiments exploring tuberculosis detection with SPAMS and results from a SPAMS field study at the San Francisco International Airport that investigated aerosol backgrounds.

Experimental

SPAMS System and Associated Hardware

SPAMS System Overview

Figure 1a shows a photograph of the SPAMS system prototype (SPAMS 1.3) that was used to produce most of the results described here. The 1.x generation of SPAMS prototypes was based on a commercial aerosol mass spectrometry system (Model 3800 ATOFMS, TSI Inc.), but had the majority of components (except mechanical frame, pumps and time-of-flight mass spectrometer) replaced by LLNL custom-built components and had additional features added, such as aerodynamic lens based particle focusing, high-rate particle tracking and sizing, and fluorescence pre-screening (*23*). A second-generation SPAMS system (2.0) was built by LLNL for DARPA/DoD (Figure 1b) and is only half the size of the system 1.3 shown in Figure 1a. Another generation of even smaller, commercial SPAMS instruments is currently under development by Livermore Instruments Inc. Figure 2 shows a schematic of a SPAMS instrument and its major components.

Aerosol Inlet and Particle Focusing

For field applications, a virtual impactor (MesoSystems MVA400, MesoSystems Technology, Inc., modified for SPAMS use) is usually attached to the particle inlet to provide significantly enhanced particle concentration. For operation in the laboratory, the SPAMS inlet is often used as is without the virtual impactor (VI). This two-stage VI draws in ~400 Lpm and concentrates particles in the size range of 0.7 -10 μm into an air stream of ~1 Lpm that is sent into the SPAMS pressure flow reducer and a high-flow aerodynamic lens.

Aerodynamic lenses are commonly used in aerosol sciences to produce finely focused aerosol particle beams (*24*), but have been limited, in the past, to relatively low flow rates (~0.1 Lpm) and small particle sizes (< 1 μm). We have created a software design tool that can be used to design aerodynamic lenses focusing any particle size range of choice (*25*). Similar work has been performed and published independently by Wang, et al. (*26*). Using our tool, have designed and built aerodynamic lenses for SPAMS that can focus a large particle size range (0.7-10 μm) and can accommodate a larger flow rate by use of the pressure flow reducer (PFR). The PFR helps to match the airflow from the exit of the VI to the inlet of the aerodynamic lens without noticeable particle losses (*27*). The aerodynamic lens focuses the aerosol particles into a tightly focused particle beam and also accelerates each particle to a final velocity dependent on its aerodynamic diameter. [The aerodynamic diameter is an expression of a particle's aerodynamic behavior as if it were a perfect sphere with unit-density and diameter equal to the aerodynamic diameter. Particles with the same aerodynamic diameter have the same terminal settling velocity. The aerodynamic diameter is a useful parameter that is related to how far a particle will be transported, either in the environment or in the human respiratory system.] The particle beam produced by the aerodynamic

(a)

(b)

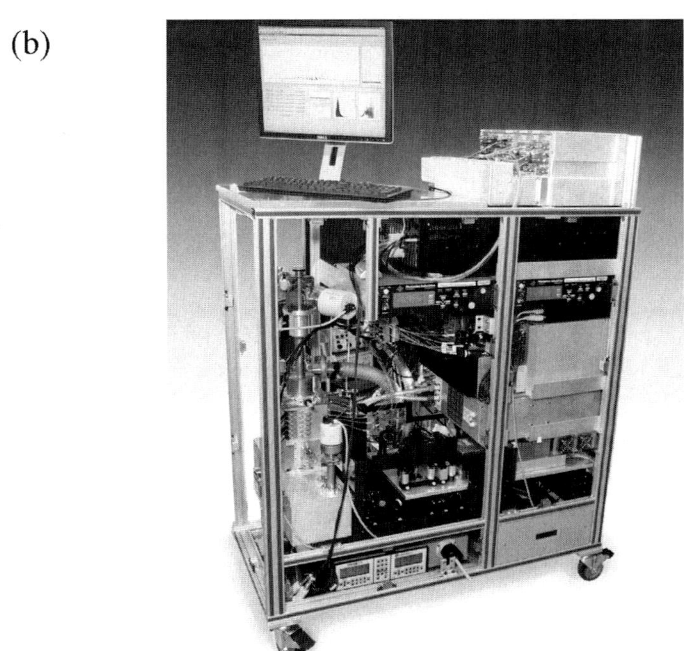

Figure 1. (a) Photo of the SPAMS system prototype (SPAMS 1.3) used to produce most of the results described here. A second generation SPAMS system (SPAMS 2.0, shown in (b)) was built for a DoD sponsor and is only half the size of the system shown (a). Even smaller SPAMS instruments are currently under development by Livermore Instruments Inc.

lens in SPAMS has a diameter of only a few 100 μm and very small divergence. Leaving the aerodynamic lens the particle beam passes through two skimmers and

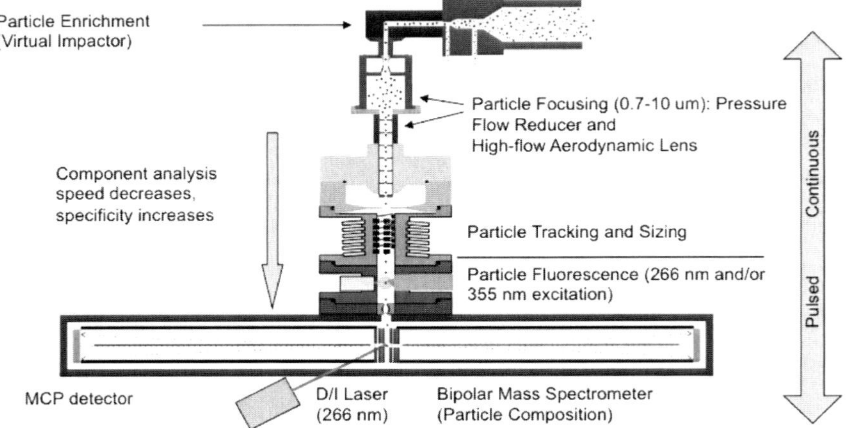

Figure 2. Schematic of a SPAMS instrument and its major components. Most LLNL SPAMS systems use a bi-polar reflectron mass spectrometer, although the use of a bi-polar linear mass spectrometer was also explored. (For simplicity, shown here is the bi-polar linear mass spectrometer.)

differential pumping stages into the particle tracking and sizing stage and continues to the fluorescence and mass spectrometry stages. Particle collection or "dusting" experiments, in which particles are captured onto a flat surface downstream from the exit of the aerodynamic lens (*28*), indicate that the particle beam diameter at the bottom of the mass spectrometer, about 30 cm from the exit of the aerodynamic lens, is less than 500 μm and corresponds to a very low divergence (~mradians) of the particle beam.

Particle Tracking and Sizing

While traversing the tracking and sizing stage aerosol particles cross six low power continuous wave red laser beams. Laser light scattered off a passing particle is detected by photomultiplier tubes (one PMT per laser beam) whose signal is conditioned and sent to an FPGA-based (field-programmable gate array) analyzer to determine particle speed. This process, as illustrated in Figure 3, allows multiple particles to be tracked at the same time. Because the speed of particles exiting the aerodynamic lens depends on the particles' aerodynamic diameter – given by a non-linear, but smooth function that can be determined through calibration with particle standards – measuring a particle's speed in the tracking stage essentially also measures its aerodynamic diameter (*29*). Experiments and simulations have shown that the SPAMS tracking stage using six lasers can track ~10,000 particles per second allowing tracking and sizing even in the highest background situations.

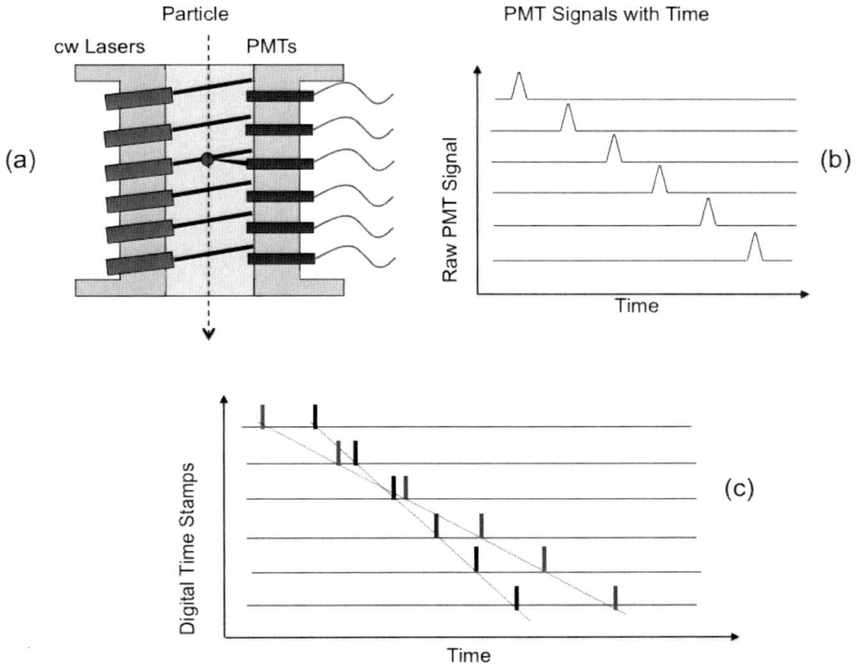

Figure 3. (a) Schematic cross section of SPAMS particle tracking and aerodynamic sizing stage equipped with six red cw diode lasers and six photomultiplier tubes (PMTs). (b) As a particle moves down through the tracking stage and crosses each of the six cw laser beams the scattered light causes subsequent signals in each of the six PMTs. (c) The PMT signals are conditioned, digitized and processed by FPGAs. Multiple particles can be tracked and sized simultaneously.

Particle Fluorescence Prescreening

After the tracking and sizing stage, particles pass through one or two (depending on SPAMS system version and configuration) particle fluorescence stages where they can be probed by low-power UV fluorescence excitation, typically of order 0.01 J/cm^2, well below the particles' desorption and ionization threshold. Each SPAMS fluorescence stage uses a pulsed, high rate (up to 30,000 pulses/second) UV laser with wavelength 266 nm or 355 nm, e.g. from Spectra Physics (Spectra Physics J40BL6 for 266 nm) or Photonics Industries (Photonics Industries DC50-266 or DC50-355) that is triggered using timing information from the particle tracking stage to hit an incoming particle as it passes through the focus of an ellipsoidal mirror. Both fluorescence light and scattered 266 nm or 355 nm excitation light are collected by the mirror with high efficiency (>80% for particles within +/- 150 μm of the mirror's focus) and directed towards two PMTs. A combination of dichroic beam splitters and filters is used to define the two wavelength bands transmitted to the PMTs, as illustrated in Figure 4. The wavelength bands are chosen to capture fluorescence from characteristic biomolecules. For 266 nm excitation, for example, the short wavelength band is typically chosen from ~300-400 nm to capture fluorescence from aromatic amino acids, the long wavelength band is chosen from ~400-600 nm to capture fluorescence from compounds such as NADH and flavins (*30*).

The scattered light at the excitation wavelength is blocked out in this scheme. For our present SPAMS systems the measured fluorescence emission is not normalized to elastic scatter – as is commonly done in solely fluorescence based particle detectors. (Normalizing fluorescence to scatter in SPAMS could be added with straightforward hardware and electronics modifications, if desired.) In SPAMS, particle fluorescence measurement is mainly used as a prescreening tool in high-background situations to select a small subset of particles that will be interrogated further by mass spectrometry. Most biological particles are presumed to be fluorescent. Thus, detecting certain fluorescence behavior from a particle passing through the SPAMS fluorescence stage can be used to decide whether that particle should be interrogated further by mass spectrometry or not. The SPAMS mass spectrometer operates at much lower rate than the tracking and fluorescence stages – tens per second vs. tens of thousand per second, respectively. Therefore pre-selecting potentially biological particles (as determined by fluorescence) for mass spectrometry will increase the sensitivity of the SPAMS system for biological aerosols when significant concentrations of non-biological (not properly fluorescent) particles are present. A similar concept was also implemented and explored, independently from our work, in an aerosol mass spectrometer developed by a group at TNO (*31*). In all SPAMS systems, prescreening can be switched on or off on the fly, as desired. While fluorescence measurements are primarily used for particle pre-screening, fluorescence information from individual particles may also be used, in some cases, in a probabilistic way to aide particle identification and, ultimately, the triggering of alarms (positive detection of target aerosols).

Laser Desorption/Ionization and Mass Spectrometer

After passing the fluorescence stage(s), particles continue on their downward path and reach the center of the ion source region of the bipolar mass spectrometer, which is at electrical ground potential. Triggered by timing information from the tracking stage and, if pre-screening is enabled, a "fire-at-this-one" signal from the fluorescence stage(s) a desorption/ionization (D/I) laser pulse from a nanosecond 266 nm Nd:YAG laser (Big Sky Ultra, for most generations of SPAMS instruments) is fired at a selected particle. The D/I laser pulse desorbs and ionizes atoms and characteristic molecules from the particle. Following results from our early studies on the laser power and profile dependence of SPAMS mass spectral variability (*32*), described in more detail below, we have added simple optical components to the D/I laser light path for all of our SPAMS systems that produce a flat-top laser profile at the center of the mass spectrometer ion source region (*33*). [For a flat-top laser profile, all aerosol particles hit by the laser are exposed to practically the same fluence, whereas for the more common Gaussian laser profile the fluence a particle is exposed to will depend on its location in the laser beam if the width of the Gaussian beam is comparable to or smaller than the diameter of the particle beam entering the SPAMS mass spectrometer ion source region.] Positive and negative ions produced by the laser pulse from a particle are accelerated in opposite directions into two back-to-back time-of-flight mass spectrometers, one for each polarity, towards micro-channel plate detectors (MCPs).

SPAMS systems of generations 1.x use the original TSI bipolar reflectron mass spectrometers (from TSI Model 3800 ATOFMS) that employ a two-step acceleration over a few cm to ~6 keV final ion energy. SPAMS 2.0 uses a LLNL custom-built miniaturized mass spectrometer that offers comparable performance at only ~1/5 of the size of a TSI mass spectrometer.

We have also explored the use of a linear mass spectrometer with guide wire to increase sensitivity and mass range (*34*). While we did observe some increase in sensitivity and mass range in this linear configuration there was a tradeoff in mass resolution. Because most of the mass spectral signatures from bacteria or bacterial spores observed with SPAMS (using plain 266 nm laser desorption/ionization) were in the mass range below ~500 Da and because that mass range was easily accessible with superior mass resolution in the original reflectron configuration, this modified, linear mass spectrometer configuration was not pursued further for SPAMS biodetection applications.

Another approach that was explored to increase sensitivity and mass range of SPAMS was aerosol matrix-assisted laser desorption/ionization (aerosol MALDI). In aerosol MALDI mass spectrometry, aerosol particles coated with a matrix are either produced from a particle sample suspended in a MALDI matrix solution that is aerosolized or aerosol particles are coated on-the-fly with a vaporized MALDI matrix. Others have explored this approach before obtaining some encouraging results but have also reported challenges in obtaining reproducible mass spectra with aerosol MALDI-MS presumably due to non-uniform matrix coating of aerosol particles (*35–39*). Using aerosol MALDI, where peptides were mixed with a MALDI matrix (2,5-Dihydroxybenzoic acid or 2,5 DHB) and aerosolized, we could demonstrate the very high sensitivity of the SPAMS method reaching a 14 zmole detection limit for Gramicidin S, a peptide with molecular mass of ~1141 Da (*40*). SPAMS of micron-sized particles containing polyethylene glycol (PEG) and various MALDI matrices were also explored (*41*). The matrix 2,5 DHB generated singly charged PEG ions of *m/z* up to ~4000. A derivatized PEG, poly(ethylene glycol) 4-nonylphenyl 3-sulfopropyl ether, generated both positive and negative ions to *m/z* = 1500. We also performed further investigations on various parameters contributing to the efficient ion generation in aerosol MALDI mass spectrometry using SPAMS (*42*). In spite of some encouraging results, the aerosol MALDI approach was not pursued further by us for SPAMS biodetection applications, mostly for practical reasons (including keeping biodetection by SPAMS a truly reagentless method and maintaining a high level of SPAMS signature reproducibility).

We also explored the use of other desorption/ionization laser wavelengths, including, in particular, 355 nm. While that wavelength did not produce feature-

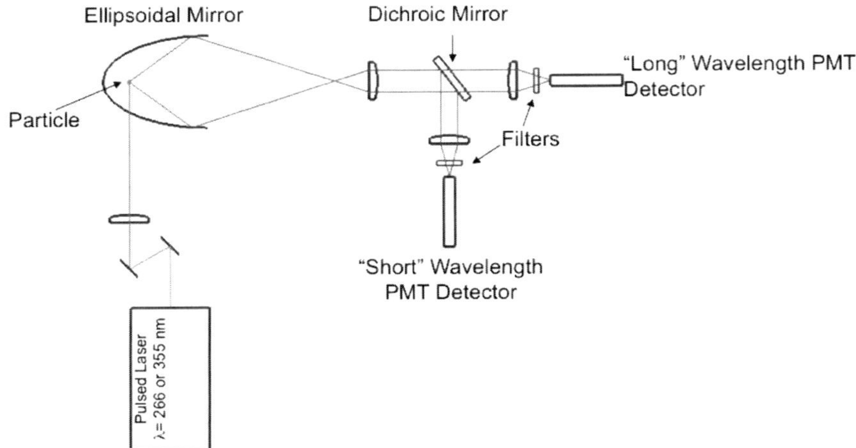

Figure 4. Schematic of a SPAMS single-particle fluorescence stage. As a particle (moving into the plane) crosses though the focus of an ellipsoidal mirror it is illuminated by a nanosecond light pulse from a 266nm or 355nm UV fluorescence excitation laser (triggered by timing information from the tracking stage above). Fluorescent light emitted from the particle is collected by the mirror and sent towards two PMT-based detectors. The wavelength bands transmitted to each of the detectors can be tailored with a combination of a dichroic mirror and filters.

rich spectra from bacteria or bacterial spores that would be comparable to spectra generated with 266 nm laser pulses it appeared to work reasonably well for some aerosolized organic compounds (*43*).

Data Acquisition and Processing

Positive and negative ion SPAMS time-of-flight spectra are produced by recording the signals from the two corresponding micro channel plate (MCP) detectors that are fed straight into two digitizer boards (PDA1000 from Signatec Incorporated for most SPAMS versions) without amplification. The boards simultaneously record time-of-flight spectra of thirty to a hundred thousand points for each polarity. The recorded time-of-flight data are converted into calibrated mass spectra using empirically derived formulas (containing an essentially quadratic dependence of *m/z* on TOF with a small offset and higher-order correction term) relating flight time to mass-to-charge ratio. Mass spectral calibrations, which take a few minutes, are easily performed using aerosol particles produced using chemical standards such as CsI, an Arginine-DPA mixture or polyethylene glycol derivatives (*41*) and can simply be performed in the field. Usually, external calibration is used, i.e., the calibrant is aerosolized separately from the sample to be measured. If desired, internal calibration – mixing the calibrant with the sample to be aerosolized and measured in the lab – can be performed as well. In the field, instrument calibration is performed regularly, typically along with inlet cleaning every week or two; total maintenance taking ~30 min.

Note that the calibration is performed using averaged spectra and that there may be a slight shot-to-shot (particle-to-particle) variation of characteristic time-of-flight peaks in the SPAMS spectra of nominally similar aerosol particles. This effect is, to a large part, due to the slight spatial spread of aerosol particles entering the SPAMS ions source region. Due to particle beam divergence particles are spread out over ~300 μm when they reach that region. As a result of this spatial spread, subsequent particles may be hit by the D/I laser in slightly different locations in the ion source (with slightly different electrical potential). Thus, ions produced from subsequent nominally similar particles may experience slightly different accelerating potentials and produce time-of-flight peak positions that vary slightly from particle to particle (typically corresponding to ±0.5 mass units or less). We have developed an autocalibration routine that will automatically correct these small shot-to-shot fluctuations before spectra are fed into pattern matching algorithms. The autocalibration routine assumes that mass spectral peaks for $m/z < 200$ fall nearly onto integer values – a reasonable assumption for organic molecules comprised of C, N, O and H – and shifts peaks that are off from integer m/z towards the nearest integer m/z, accordingly. Another calibration algorithm developed and used, on occasion, searched for common peaks in the mass spectra from a number of particles, as defined by the user, and calibrated the spectra based on any that it found. Although perhaps not as widely applicable as the autocalibration algorithm mentioned earlier, this algorithm does not require any assumption about integer mass.

Identifying individual aerosol particles based on the measured SPAMS mass spectra is accomplished by pattern recognition that may be complemented by the application of mass spectral rules trees (*22*). For automated particle identification, we have developed a real-time pattern recognition algorithm that has been implemented on a nine digital signal processor (DSP) system from Signatec Incorporated (*44*). The algorithm first pre-processes the raw time-of-flight data through an adaptive baseline removal routine. The next step consists of a polarity-dependent calibration to a mass-to-charge representation, reducing the data to about three hundred to a thousand channels per polarity. The next step is rough identification using a pattern recognition algorithm based on a dual-polarity library of known particle signatures including threat agents and background particles. The final identification step includes a score-based rule tree. This algorithm has been implemented on the PMP8A from Signatec Incorporated, which is a computer based board that can interface directly to the two one-Giga-sample digitizers (PDA1000 from Signatec Incorporated) used to record the two polarities of time-of-flight data. By using optimized data separation, pipelining, and parallel processing across the nine DSPs it is possible to achieve a processing speed of up to a thousand particles per second, while maintaining the recognition rate observed on a non-real time implementation.

In practice, the rate at which SPAMS can analyze and identify particles by mass spectrometry is limited by the rate with which the D/I laser can operate. The present SPAMS D/I lasers are based on flash lamp pumped Nd:YAG lasers with rates of 20-50 pulses per second (depending on model) for 266 nm wavelength and milliJoule pulse energies. The overall particle hit efficiency (i.e. percentage of tracked particles that are fired upon and actually hit by the D/I laser to produce mass spectra) depends on particle size and type, but is generally between 10 and 50 %. Thus the overall acquisition rate of mass spectra from individual particles is typically ~5-20 per second (assuming sufficient aerosol concentration at the inlet). Higher rate D/I lasers with mJ energy and pulse rates of order kHz are commercially available (but somewhat bulky), and could be used if higher acquisition rates were desired. Even with the currently used relatively slow D/I lasers, given the relatively high efficiencies of inlet, aerosol focusing and tracking together with fluorescence screening SPAMS is able to achieve remarkable detection sensitivities in the range of one to some tens of ACPLA ("agent containing particles per liter of air") – depending, of course, on threat aerosol type, backgrounds and acceptable false alarm rate – for detection in one minute or less.

Aerosol Generation in the Laboratory

Aerosols have been generated in numerous ways for study in the laboratory. For dry powders, such as lyphoyzed spores or inorganic powders, we have used 1-L 0.22 μm filter/sterilizers (Corning, Inc., Corning, NY) where, upon agitation, aerosolized powder can be drawn out into tubing by the SPAMS vacuum for analysis. For liquid suspensions of biologicals or other material, or solutions of chemicals we used numerous types of Collison nebulizers for aerosolization using a flow of nitrogen gas. A single-jet nebulizer (TSI Inc., St. Paul, MN) or a six-jet modified MRE-type Collison nebulizer (BGI, Inc., Waltham, MA) can be used for prolonged aerosolization due their large volume. A convenient way to introduce aerosols of suspensions or solutions of samples into SPAMS is using a low volume (2-5 mL) disposable nebulizer (e.g., from Salter Labs, Arvin, CA, USA) (*41*), where each sample can have its own nebulizer without fear of cross contamination and can be stored for later use. Solutions are aerosolized by introducing a flow of 1.5 Lpm nitrogen gas into the Salter nebulizer. In all nebulization methods used for SPAMS measurements, the aerosol droplets are dried through a desiccant column and the particles are sampled into the SPAMS inlet through copper or conductive silicone tubing. In order to be able to more accurately control the size of generated particles, we have used a Sono-tek ultrasonic nozzle (Sono-Tek Corp, Milton NY) for microbial suspensions and a Vibrating Orifice Aerosol Generator (VOAG, TSI Inc., St. Paul, MN) for chemical solutions.

Biological Sample Preparation

Detailed descriptions of the biological samples used and their preparations for SPAMS measurements can be found in the literature cited throughout the experimental section below. For *Bacillus* spores, one of the most prominent sample types studied with SPAMS, a brief description is given here as well. *Bacillus atrophaeus*, formerly *B. globigii*, is frequently used as a surrogate for *Bacillus anthracis* by the Department of Defense. For most of the experiments described here, *B. atrophaeus* cells (ATCC #9372, Dugway Proving Ground, Dugway, UT) were grown to mid-log phase in tryptone yeast extract broth or nutrient broth and then aliquoted into ¼ TY media in a 1:25 or 1:50 dilution.

The cells sporulated in a shaker incubator at 32°C until approximately 90% were refractile (3-4 days). Phase contrast microscopy and spore staining confirmed that spores were in fact present. The spores were then harvested by centrifugation at 8000g for 12 minutes, and washed in cold double-distilled water. After three washes the spores were reconstituted in double distilled-water at concentrations of $\sim 10^7$ spores/mL, as determined using a Petroff-Hauser counting chamber (Hausser Scientific Partnership, Horsham, PA). Using a similar procedure, *Bacillus thuringiensis* (ATCC No. 16494) spores were also prepared and included as a surrogate for naturally occurring particle types that might easily be misidentified as *B. anthracis* (or *B. atrophaeus* as used in the experiments described here).

Results and Discussion

Bacillus Spores: Typical Mass Spectral Signatures and Species Discrimination

Figure 5 shows results from a typical SPAMS measurement of aerosolized bacterial spores. For this measurement performed in the laboratory, spores of *Bacillus atrophaeus* (formerly *B. globigii*, a common simulant for *B. anthracis*) were aerosolized out of aqueous solution with a simple nebulizer (Collison nebulizer), passed through a diffusion dryer and introduced into the front end of the SPAMS system (without virtual impactor). The measured size distribution of those spore particles – as determined by aerodynamic sizing in the SPAMS tracking and sizing stage – is plotted in Figure 5 (a). The size distribution shows a relatively narrow peak at around 1.1 μm, as expected for single *Bacillus* spores (*45*) indicating that most particles passing through this stage (and on to the fluorescence and mass spectrometer stages below) are, indeed, individual spores.

Figure 5 (b) shows the measured particle fluorescence using pulsed 266 nm laser excitation and detection in two emission wavelength bands, from 290-400 nm and from 400-600 nm. Each dot in this scatter plot corresponds to the measurement of an individual particle. The gain for the two fluorescence channels (as set by the PMT voltage for each channel) is generally chosen such that the data points from the spores make optimal use of the full digitizer range. The fluorescence signal for spores fall roughly on a diagonal line in the scatter plot. In our SPAMS systems, the measured fluorescence emission is not normalized to elastic scatter and individual particles may be exposed to different excitation fluences depending on where exactly in the UV laser beam they are located when excitation happens. Therefore, the data points from similarly composed (i.e., similarly fluorescing) particles will scatter along a line and not be confined to a single "blob". This diagonal region in the plot (above some threshold near the origin) can, therefore, be defined as a potential "spore threat region". Typical backgrounds from indoor or outdoor environments generally fall below this "spore threat" region, as indicated in this figure. Other types of biological materials (vegetative cells, viruses, toxin simulants) can be found in other regions of this plot (not shown here). For actual, practical detection applications, the SPAMS real-time detection and alarm software allows the user to define "threat" regions that are much more complex and structured than the simple oval shown

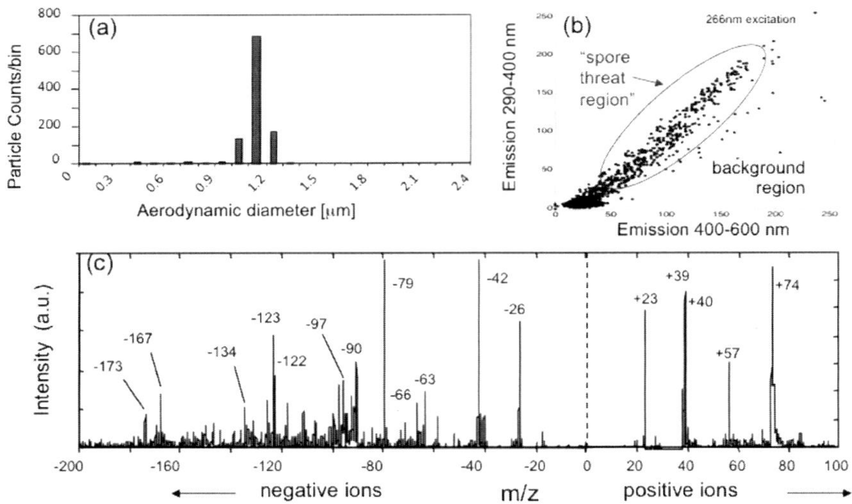

Figure 5. Typical experimental results from measuring aerosolized Bacillus atrophaeus spores with SPAMS. (a) Bacillus particle size distribution (as determined by aerodynamic sizing in the SPAMS tracking stage) peaking at 1.1 um, the size of individual spores; (b) measured particle fluorescence using 266 nm laser excitation and detection in two emission wavelength bands. Each dot in this scatter plot corresponds to the measurement of an individual particle. (c) measured dual-polarity mass spectrum of a single B. atrophaeus spore. This spectrum was produced by concatenating the two measured mass spectra for the positive and negative ions from the same spore particle (indicating the negative ions with a negative m/z).

in this figure. It should also be noted that identification of a particle as a potential "threat" by the SPAMS fluorescence prescreening stage(s) generally only guides the system to investigate such particles further with mass spectrometry and does not produce an alarm, or positive detection, by itself.

Figure 5 (c) shows the measured dual-polarity mass spectrum of a single *B. atrophaeus* spore. Since both polarities of ions are measured simultaneously from the same particle it is convenient and instructive to combine the corresponding positive and negative ion mass spectra into a single spectrum (with both positive and negative *m/z*), as done here, although, technically, they are measured by separate, back-to-back mass spectrometers. For the pattern analysis, the user can choose to perform this analysis on the concatenated spectra or the positive and negative ion spectra separately.

To aide *m/z* peak identification (i.e., assigning specific chemical species to the prominently observed mass spectral peaks) we performed isotope labeling experiments comparing mass spectra from *Bacillus* vegetative cells and spores that were grown in ^{15}N-labeled and ^{13}C-labeled growth media to spectra from cells and spores grown in unlabeled media (*46*, *47*). Figure 6 shows some results of isotope labeling experiments from (*47*) and peak assignments in the spore mass spectra derived from the results of such experiments.

In addition to the isotope labeling experiments, we compared vegetative cell and spore mass spectral signatures with mass spectral signatures from pure chemical standards and mixtures of standards compounds known to be present in spores. Effects of the compounds' acid-base gas-phase chemistry (*48*) and effects from aromatic ring containing nucleobases (*47*) were studied with SPAMS as well and helped explain the observed signatures. Lastly, information from literature on bacterial spore composition and known bacterial metabolic pathways was used to check that our peak assignments made sense. For example, prominent spore mass spectral peaks at *m/z* -134, -146, -167 and -173 were assigned to aspartic acid, $[Asp-H]^-$, glutamic acid, $[Glu-H]^-$, dipicolinic acid, $[DPA]^-$, and arginine, $[Arg-H]^-$, respectively. These compounds are known to be present at significant concentrations in *Bacillus* spores and can be linked to known metabolic pathways (e.g., Aspartic acid, Glutamic acid and DPA are compounds found in the aspartate pathway of *Bacillus*).

Figure 7 shows typical mass spectra from *B. atrophaeus* spores (top) and *B. thuringiensis* spores (bottom). Overall, the spectra from the two species are similar. However, two characteristic marker peaks (*m/z* -173 and *m/z* +74 indicated by the ovals) are clearly present in the *B. atrophaeus* spectrum and absent (or very weak) in the *B. thuringiensis* spectrum, allowing the two species to be discriminated from each other by SPAMS (*12*). (The finding that these two species can be discriminated based on those two markers peaks was purely empirical and has been confirmed in numerous subsequent SPAMS experiments.) To improve visibility of characteristic peaks average spectra are shown here (average of 1000 single-particle mass spectra). Nevertheless, the discrimination between these two species works with high confidence (>90%) with single spore spectra when the SPAMS pattern matching and rules algorithms are used.

Our work on bacterial spore species discrimination, based on single spore mass spectra measured with SPAMS, has been expanded to include 10 different *Bacillus* spore species (including *Bacillus anthracis* Sterne) and *Clostridia* spores. We found that *Clostridia* spores can be discriminated from *Bacillus* spores by SPAMS. Also several of the *Bacillus* spore species can be discriminated from each other, in the same way as the two species shown in Figure 7. However, in these studies, some closely related *Bacillus* spore species yielded SPAMS mass spectral signatures that were too similar to allow species discrimination with high confidence based on single spore mass spectra alone.

SPAMS Measurements on Vegetative Bacterial Cells

Our SPAMS work with vegetative bacterial cells included vegetative cells of several *Bacillus* species (some of the same species the above mentioned spores were produced from), *Enterobacter agglomerans* (also called *Erwinia herbicola* or *Pantoea agglomerans*; a *Yersinia pestis* simulant), some non-pathogenic strains of *Escherichia coli*, some mycobacterial species (discussed further below) and others. As an example, Figure 8 shows a typical SPAMS mass spectrum from *Bacillus atrophaeus* vegetative cells. To improve visibility of the characteristic mass spectral peaks in this figure a ~1000 shot average of single cell mass spectra is shown here. It should be noted, though, that SPAMS pattern recognition and

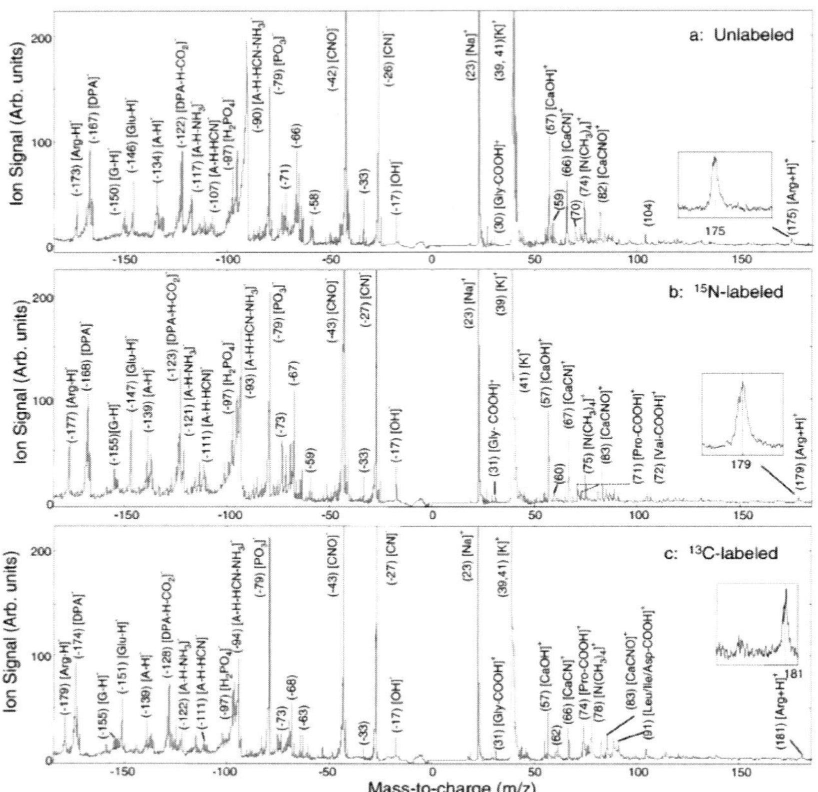

Figure 6. Typical Dual polarity average mass spectra of B. atrophaeus spores (a-c: 324, 420, and 125 spectra averaged, respectively) grown in unlabeled (a), ^{15}N-labeled (b), and ^{13}C-labeled (c) Bioexpress growth media and peak assignments. The inset in each panel shows the protonated arginine peak (from ref. (47)).

rules based algorithms are always applied to single-particle spectra (and not to averages) and can identify single-particle spectra from bacterial vegetative cells with fairly high confidence (~70-80%). Ion peaks from *B. atrophaeus* vegetative cells in the negative polarity are attributed mostly to phosphate clusters, ion peaks in the positive polarity are attributed mostly to amino acid residues.

In our work with vegetative bacterial cells we found that SPAMS can discriminate on single shot (single cell) basis between several of the bacterial species investigated. In particular, discrimination between some of the relevant biological agent simulant species (e.g., *E. herbicola*) and potential "interferents" (e.g. *E. coli*) was demonstrated. For *Bacillus* vegetative cells, on the other hand, we found that the vegetative cell SPAMS mass spectra of the various *Bacillus* species we studied are more similar to each other than the mass spectra from the corresponding *Bacillus* spores were to each other. Consequently, we have not been able to obtain consistent species discrimination with SPAMS between vegetative cells of various *Bacillus* species comparable to the species discrimination we

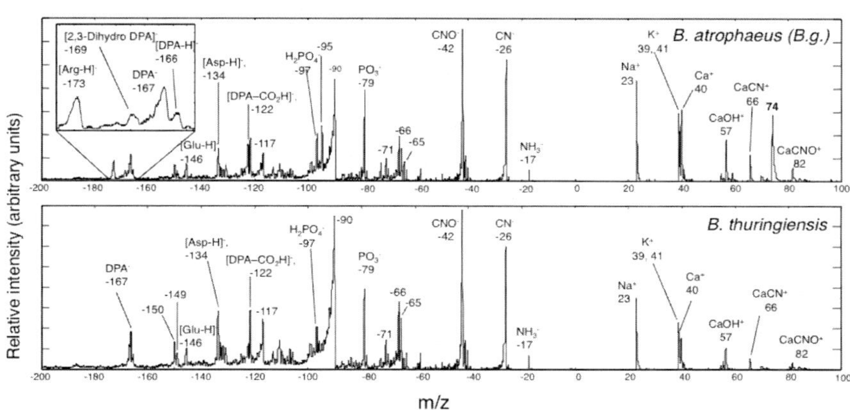

Figure 7. Typical mass spectra from B. atrophaeus spores (top) and B. thuringiensis spores (bottom). Two characteristic marker peaks (m/z -173 and m/z +74 indicated by the ovals) are clearly present in the B. atrophaeus spectrum and absent (or very weak) in the B. thuringiensis spectrum, allowing the two species to be discriminated from each other by SPAMS. To improve visibility of characteristic peaks average spectra are shown here (1000 particle mass spectra averages) However, the species discrimination between these two species works with high confidence (>95%) with single spore spectra when the SPAMS pattern matching and rules algorithms are used.

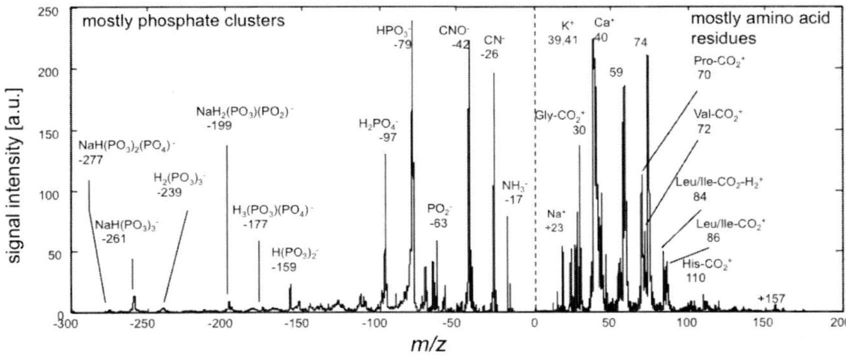

Figure 8. Typical SPAMS mass spectrum from Bacillus atrophaeus vegetative cells. To improve visibility of the characteristic mass spectral peaks in this figure an average of 1000 single cell mass spectra is shown here. Ion peaks in the negative polarity are mostly assigned to phosphate clusters, ion peaks in the positive polarity are mostly assigned to amino acid residues.

could show for spores of the corresponding species (when confined to using only single-particle mass spectra for the discimination).

Studying Consistency of Bacterial Signatures as Function of Instrumental Conditions, Growth Conditions, Sample Preparation and Aerosolization Methods

For biodetection applications, it is not only important to understand the origin of observed mass spectral marker peaks for the various classes and species of biological aerosols of interest, but also to understand the robustness (or variability) of such marker peaks as a function of instrumental conditions, growth conditions, sample preparation and aerosolization methods. Over the last years, we have performed systematic studies with SPAMS in these areas for a large variety of biological agent simulants including bacteria and bacterial spores, viruses and toxin simulants. Some of this work and resulting findings are summarized briefly in the following.

Effects of Laser Power and Profile on Observed SPAMS Mass Spectra from Bacteria

As in other types of nanosecond laser desorption/ionization based mass spectrometry (LDI-MS, MALDI-MS), the ion signal and fragmentation observed in the mass spectra measured by SPAMS depend very much on the laser fluence (energy per area) used. [For lasers with pulse lengths of nanoseconds pulse length and pulse energies in the milliJoule range focused to spot sizes of order 100 μm diameter, multiphoton absorption processes should be negligible and the total energy absorbed by a particle as determined by the fluence (measured in J/cm^2) is the relevant parameter for describing desorption and ionization. For short-pulse ("femtosecond") lasers, multi-photon processes are more important and the intensity (measured in W/cm^2) is, likely, the more relevant parameter in that case.] In one of our early studies with SPAMS and *Bacillus* spores, the fluence dependence of the observed mass spectra was explored (*32*). Based on the observations from this work, we implemented a flat-top laser profile in the SPAMS systems to improve the mass spectral consistency for biological aerosols and to further study the desorption and ionization (D/I) fluence threshold (*33*). In these studies, we found that the D/I threshold for Bacillus spores was around ~0.1 J/cm^2 and that fluences around ~0.2 J/cm^2 resulted in the "optimal" SPAMS mass spectra from *Bacillus* spores ("optimal" meaning least fragmentation and largest signal at larger m/z). We also found that the optimal fluence for obtaining good SPAMS mass spectra from vegetative bacterial cells, virus samples and toxin simulant preparations was higher, by about a factor 3-4, compared to the optimal fluence for spores. The relatively lower optimal fluence for spores maybe explained by the fact that *Bacillus* spores contain about 10% DPA, which can act as an internal matrix and facilitate desorption/ionization similar to MALDI-MS, although no external matrix is used here. This hypothesis is supported by experimental results from our SPAMS studies on mixtures of DPA with various amino acids and other compounds (*48*).

The ideal fluence for producing information-rich SPAMS mass spectra from biological aerosol particles also depends somewhat, but to a much lesser extent, on particle size, which in turn depends on sample preparation and aerosolization method used. For example, using a Collison nebulizer to aerosolize *Bacillus* spores from an aqueous solution at a concentration of ~-10^7-10^8/mL will generate predominately particles with an aerodynamic diameter around 1 μm (containing only a single spore) whereas using a Sono-Tek atomizer as the aerosol generator creates particles predominantly in the ~3 μm aerodynamic diameter range (containing multiple spores). It should be noted, though, that the mass spectral signal size (peak area) observed with SPAMS does not scale linearly with particle volume, indicating that not all of the material in a particle is vaporized and ionized when the "optimal" laser fluence is used.

In practical biodetection applications of SPAMS it would not be known, a priori, what type of agent the system might be challenged with. Based on the results described above, the system would, therefore, be operated at a fluence that will detect the broadest range of biological threats with high efficiency (but might not be optimal for species discrimination) and, if bacterial spores were detected, the fluence would quickly be lowered (by changing the laser pulse energy) to investigate the exact species of the spores further. In fact, the newer generation of SPAMS system performs this operation automatically based on observed signatures.

Even after implementing the flat topped D/I laser profile, some shot-to-shot variability among the mass spectra from nominally similar biological particles was observed (*33*). Further investigation could attribute part of this variation to ion statistics: except for the strongest peaks in the SPAMS mass spectra from biological aerosols, most peaks were found to be produced by only a few tens of ions impacting onto the MCP detectors. Poissonian statistics of the number of these ions can cause a noticeable shot-to-shot variation in ion number (and corresponding peak height) for a given *m/z* (*49*). However, this study also showed that some part of the shot-to-shot variation of SPAMS mass spectra from nominally similar biological particles cannot be explained by the combination of residual variations in laser fluence (remaining after the flat topping) and statistical fluctuations alone. Presumably, this remaining variation is due to true biological (composition and shape) variations between individual particles of the same kind and will be hard to overcome by further instrumental improvements. Practical detection and alarm algorithms will need to be flexible enough to deal with this remaining variation in the measured mass spectra of nominally similar biological particles. (In the case of SPAMS, this is accomplished, e.g., by having several, slightly different, mass spectral patterns in the database for a given type of agent simulant.)

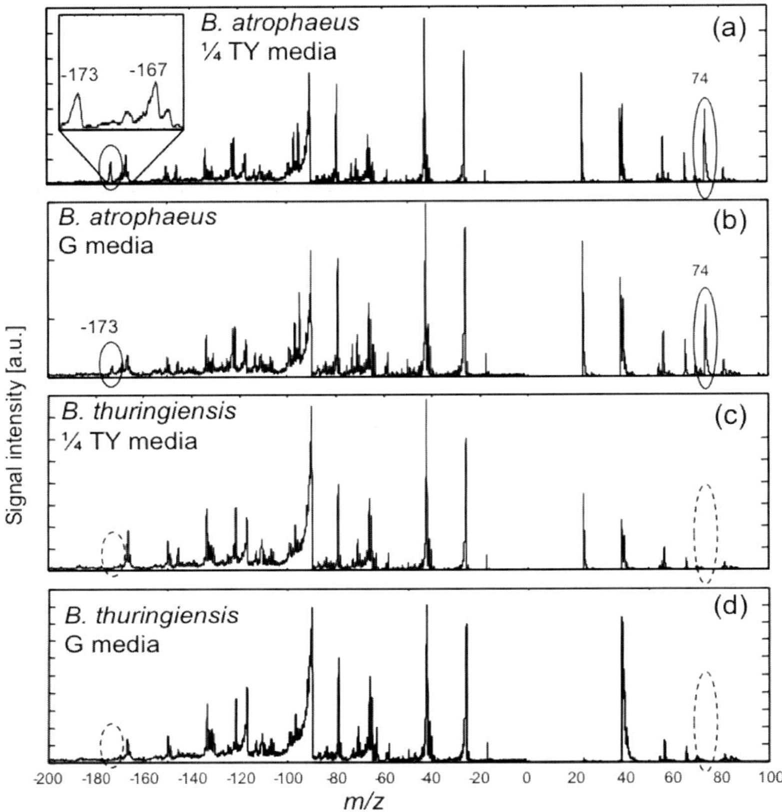

Figure 9. Typical SPAMS mass spectra from B. atrophaeus spores (a, b) and B. thuringiensis spores (c,d) grown in ¼ TY media (a,c) and G media (b,d), respectively. The peaks at m/z +74 (betaine fragment) and -173 (Arg) help discriminate B. atropheaus (both peaks clearly present for both media) from B. thuringiensis (both peaks absent or very weak for both media).

Effects of Growth Conditions and Media on Observed SPAMS Mass Spectra from Bacteria

We also have performed various studies on the dependence of SPAMS mass spectral signatures from bacteria and bacterial spores on growth media used and growth conditions applied. Fundamentally, growth media could affect mass spectra from organisms grown in the media in two ways: 1) media residues could be found on (or in) the microorganisms even after cleaning and washing, and 2) the type of media used to grow a particular microorganism could influence the metabolites or composition and concentration of other compounds found in the microorganisms.

Figure 9 shows the results from an experiment in which *B. atrophaeus* and *B. thuringiensis* vegetative cells were grown and sporulation was induced in two different media, ¼ x TY (nutrient-rich) and G-media (minimal media), before purification and resuspending the spores in water. The SPAMS mass spectra shown in this figure are from spores resulting from this procedure. All of the four spectra shown appear similar (just from purely visual inspection) and are as expected for *Bacillus* spores, indicating that the overall spore spectra likeness (e.g, characteristic DPA peak at m/z -167) of the spectra is preserved between the two Bacillus species and when ¼ TY media is replace by G media. The peaks at *m/z* = +74 (betaine fragment) and *m/z* = -173 (Arg) help discriminate *B. atropheaus* (both peaks clearly present for both media, Figure 9 a and b) from *B. thuringiensis* (both peaks absent or very weak for both media, Figure 9 c and d) are also preserved between the two media.

The overall conclusion from this work was that the media dependence on observed SPAMS signatures is generally small (if not unnoticeable) for the *Bacillus* species and growth media studied. However, our studies on growth media dependence of mass spectral signatures were certainly not comprehensive, and cannot be used to draw broader conclusions on media dependence or independence of SPAMS mass spectral signatures. In contrast to this work with bacterial spores, when studying *Mycobacterium tuberculosis* and *M. smegmatis* with SPAMS we did observe mass spectral peaks that were correlated with bacterial growth media (see below). In addition, work by others using different types of mass spectrometry has shown that there can be a media dependence of observed mass spectral peaks (*50–52*).

Not only growth media, but also other factors such as growth conditions could potentially have an impact on the composition and thus the mass spectral signatures from organisms of interest. In this context, one interesting finding for the *Bacillus* spores studied by us with SPAMS was the following observation. The prominent peak at *m/z* = +74 that helps discriminate *B. atrophaeus* (peak present) from *B. thuringiensis* (peak normally absent) can be induced for *B. thuringiensis* when *B. thuringiensis* is grown in very high salt concentrations. This *m/z* = 74 peak is attributed to a fragment of betaine, which presumably functions as an osmoprotectant for bacteria and appears in the *B. thuringiensis* spectra when the bacteria are grown in 500 mM or higher NaCl concentration (*53*). Even in this case, species discrimination between *B. atrophaeus* and *B. thuringiensis* based on single spore SPAMS spectra can still work (using, e.g, the *m/z* = -173 Arg peak that seems unaffected by the salt), albeit at a somewhat lower discrimination efficiency.

In another study, the changes occurring in single cell mass spectra during the sporulation of Bacillus cells were measured with SPAMS (*13*). *B. atrophaeus* vegetative cells were induced to sporulate by starvation. Samples were taken and mass spectrometry performed at various stages of the sporulation process over a 24-hour period. As observed by SPAMS, vegetative cell signatures morphed into spore-like signatures over this time course and about seven distinct types of *Bacillus* single cell mass spectra could be observed during that time ranging all the way from the "pure vegetative" type to the "pure spore" type. At any given time during that period, several of these spectral types were present concurrently indicating that not all cells sporulated at exactly the same rate. This study also involved a mutant that could not complete sporulation and would arrest in an intermediate state that did not contain DPA in the proto-spore. Consequently, for this mutant we did not observe the final stage containing the DPA peak in the mass spectra that is characteristic for *Bacillus* spores.

Effects of Aerosol Sample Preparation on Observed SPAMS Mass Spectra from Bacteria

The majority of our SPAMS experiments with bacteria and spores involved fairly clean samples that were washed multiple times to remove media and other residues and that were aerosolized from aqueous solution (spores, vegetative cells) or a buffered solution (vegetative cells). In practical biodefense applications, however, a biodetection system could also encounter biological aerosols that originate from a "dirty" preparation. For example, if the media is not washed away completely after vegetative bacteria or spores are grown and the sample solution is prepared for aerosolization, some media may remain attached to cells or spores. This could lead to composite aerosols (cells+media), which may exhibit additional peaks from media in the single cell mass spectra. Similarly, any non-volatile buffer (i.e., most practical buffers) used to stabilize and store a biological sample (cells, spores, viruses, proteins) and not removed before aerosolization is likely to be present on the resulting aerosol particles and likely going to produce additional peaks in the observed mass spectra.

More importantly, biological materials may intentionally be prepared to be aerosolized as composite particles in order to stabilize the material for storage and facilitate aerosolization and fine dispersion. This is done, for example, in the application of biological materials as pesticides. Instructive, in this context, is a detailed article on small-scale processing of microbial pesticides by the Food and Agricultural Organization of the United Nations (*54*) that describes a wide range of potential additives, sample preparation techniques and aerososolization methods.

Any type of biological aerosol that is comprised of composite particles ("biological" material + "additional" material) will likely not only exhibit mass spectral peaks from the biological material but also from the additional material in the aerosol particle mass spectra. (Also, some of the mass spectral peaks form the biological material of interest could be suppressed due to "matrix" effects.) In any case, additional material bound to biological material of interest in a bio-aerosol will, likely, pose challenges for the mass spectral identification based on pattern recognition, if the calibration or "training" data contain mass spectra from only "clean" preparations. For practical detection applications, it will, therefore, be essential to train an aerosol mass spectrometry based system not only to a number of biological materials of interest but also to a wide range of potential preparations of these biological materials. As a result, such a system will have multiple (presumably similar, but distinct) signatures for the same biological material in its recognition database. Recognition algorithms need to be made flexible enough to deal with the potential signature variations that such a system may encounter. In the case of SPAMS, we have relied on extensive system training in the laboratory and have developed robust identification algorithms that use both pattern recognition and rules as described in the next section.

Real Time Identification with SPAMS: Pattern Recognition and Rules

The identification of individual aerosol particles measured with SPAMS is performed using a two-step classification and identification procedure (*22*, *44*). Particle classification using SPAMS mass spectra can be performed using pattern recognition based on a modified ART 2a algorithm (*55*). More recently, we have developed an efficient way to determine similarity of mass spectra and to classify particles based on representing the mass spectra as vectors in a high dimensional space (e.g., 350-D where each dimension represents an m/z value between 1 and 350) and calculating the angle between the vectors for different particles (*32*, *33*). In this process, the patterns from positive and negative ion mass spectra from a particle are analyzed separately, at first, and compared to patterns from other particles or patterns in the library. The identification step includes integrating the two mass spectral polarities and using a score-based rule tree for the final identification determination. The identification procedure uses a library of known particle signatures including "threat" agents and "background" particles. The library signatures include both patterns and rules. Figure 10 shows an illustration of this identification process.

In this example pattern recognition and rules are used sequentially. First, the classification is performed that discriminates between vegetative cells from bacterial spores. Then, rules are used to help speciate the vegetative cells or the spores, respectively. It is also conceivable to use both types of procedures (and potentially also other algorithms) in parallel, which might be more advantageous for more complex aerosols (such as the composite aerosols discussed above) or to detect known "threat" organisms present in an unknown preparation. New patterns and/or rules can be derived from new data quickly (in minutes) using automated routines and can easily be added to the signatures in the library. Thus the system can learn in the field and signature updates can quickly be shared between instruments.

SPAMS can measure mass spectra from a large number of individual aerosol particles in a short time (tens of particles per second) and perform single-particle identification in real-time. We have also developed robust alarm algorithms for SPAMS that compile the vast amount of single-particle information further (also in real time for our most advanced SPAMS systems) and determine whether an alarm should be called or not. In this context, it is important to stress that an alarm is fundamentally different from a single particle identification.

An alarm is the result of the analysis of many particles and will only be called as the result of identifying multiple particles as a "threat" in a given time interval. Alarms are never called based on the identification of a single particle as an agent. For this reason, the SPAMS alarm algorithm is robust against individual particles remaining unidentified or being mis-identified, on occasion. A beneficial result of this, for example, is that the probability of a false alarm for a certain agent type, even in the middle of a cloud of a potential interferent, can be orders of magnitude less than the probability of misidentifying a single interferent particle as an agent particle.

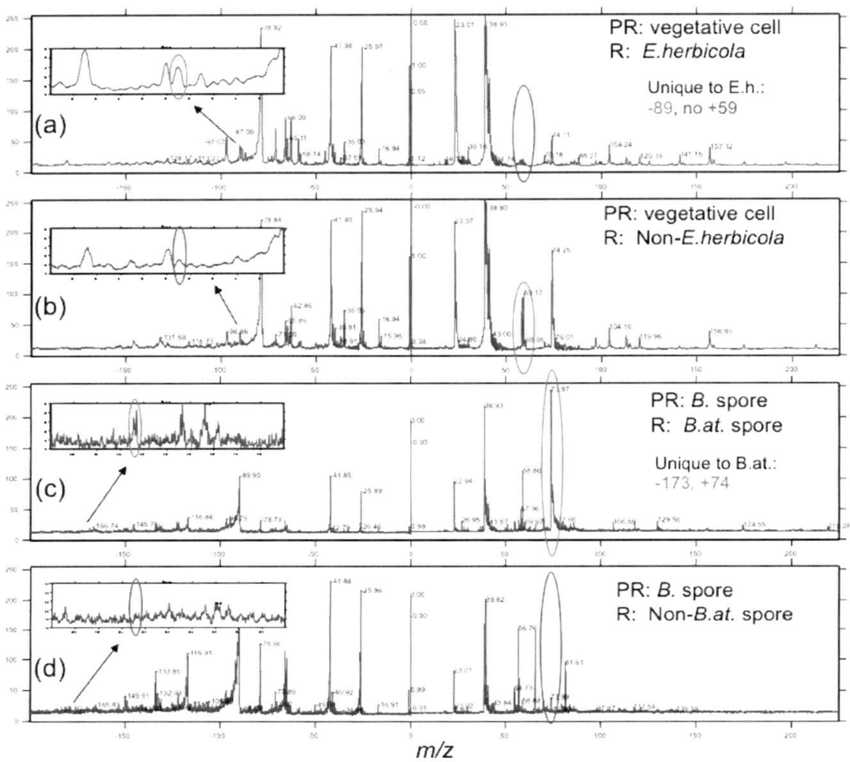

Figure 10. Illustration of pattern recognition and rules used for the identification of vegetative cells and spores. As a first step, pattern recognition is used, to provide a general classification of a particle. Pattern recognition will, e.g., recognize the spectra shown in (a) and (b) as spectra from vegetative cells and the spectra shown in (c) and (d) as spectra from Bacillus spores As a second step, rules are applied to further speciate a particle. For example, observing the presence of a peak at m/z = -89 and absence of a peak at m/z =+59 (indicated by ovals) will assign the spectrum shown in (a) to Erwinia herbicola and discriminate it from the vegetative cell spectrum in (b). Similarly, the presence of a peaks at m/z = +74 and -173 (indicated by ovals) will assign the spectrum shown in (c) to Bacillus atropheaus and discriminate it from the spore spectrum in (d).

In this context of multiple particle measurements being required for calling alarms, for practical SPAMS applications, the SPAMS system can be operated at a relatively high fluence (~1 mJ/cm^2) to be sensitive to a broad range of biological threats. When a certain type of biological aerosol is detected, the system can quickly switch to a laser fluence that is optimized for further speciation of this bioaerosol type. For example, while some speciation of bacterial spores can be obtained with SPAMS even at high fluence, the optimal speciation can be obtained at lower fluence (~0.2 mJ/cm^2), as discussed above. Therefore, if bacterial spores are detected by the pattern recognition, the system can quickly adjust the D/I laser fluence to a value more suitable for spores and continue measuring additional aerosol particles present for better speciation and to determine whether a threat is, indeed, present.

The alarm threshold can be user defined by software and is generally not a fixed threshold but dynamic. It is adjusted automatically based on the overall aerosol composition that is observed. For example, from previous training of the system it may be known that a certain type of particle is, on occasion, mis-identified as a "threat." Such a particle type would be considered an "interferent" and the SPAMS database would contain information on how often such an "interferent" particle is typically mis-identifed as threat. If SPAMS detects particles of a known type of interferent, the alarm threshold is automatically raised to account for the expected number of mis-identifications given the measured concentration of this "interferent". Thus, some "cross-talk" in the single particle identification between "threat" and non-threat aerosol species is acceptable (as long as the correct identification probability for a threat particle is reasonably high (say, >50%) and the mis-identification probability for an "interferent" as "threat" it is reasonably low, say <10%). In this case, SPAMS will still be sensitive to the "threat" and the presence of the "interferent" will lead only to a slightly higher alarm threshold, (i.e., a small reduction in system sensitivity).

Sensitivity, detection limits and false alarm rates have been thoroughly evaluated for various SPAMS systems in laboratory and field tests, but cannot be covered here in detail. From the results of those tests and side-by-side comparisons with other detection systems we can conclude that the development of SPAMS (the hardware together with the algorithms) to the present versions presents a significant and important advance in rapid aerosol threat detection.

Exploring Tuberculosis Detection with SPAMS

SPAMS was also evaluated as a potential detector for aerosols containing *Mycobacterium tuberculosis* cells in two sets of laboratory experiments that involved an avirulent Tb strain, *M. tuberculosis* H37Ra.

The first set of experiments was carried out in collaboration with CDC NIOSH in Cincinnati and investigated the potential use of SPAMS as a rapid detector for individual airborne, micron-sized, *Mycobacterium tuberculosis* H37Ra particles, comprised of a single cell or a small number of clumped cells (*14*). The SPAMS mass spectral signatures for aerosolized *M. tuberculosis* H37Ra particles were found to be distinct from *M. smegmatis* particles and vegetative cells of *B. atrophaeus*, and *B. cereus*. The discrimination used a distinct biomarker, a unique mass spectral peak at m/z = -421 that was found in the *M. tuberculosis* H37Ra spectra (see Figure 11) and hypothesized to be deprotonated trehalose-2-sulfate and likely the precursor of *M. tuberculosis* virulence factor. To our knowledge, this was the first time a potentially unique biomarker was measured in *M. tuberculosis* H37Ra by mass spectrometry on a single-cell level.

In these collaborative experiments in the CDC NIOSH labs, SPAMS was also coupled to a wind tunnel that also contained other types of aerosol sampling and reference equipment, such as an aerodynamic particle sizer (Model 3321 APS, TSI Inc.), a viable Anderson six-stage sampler, and filter cassette samplers

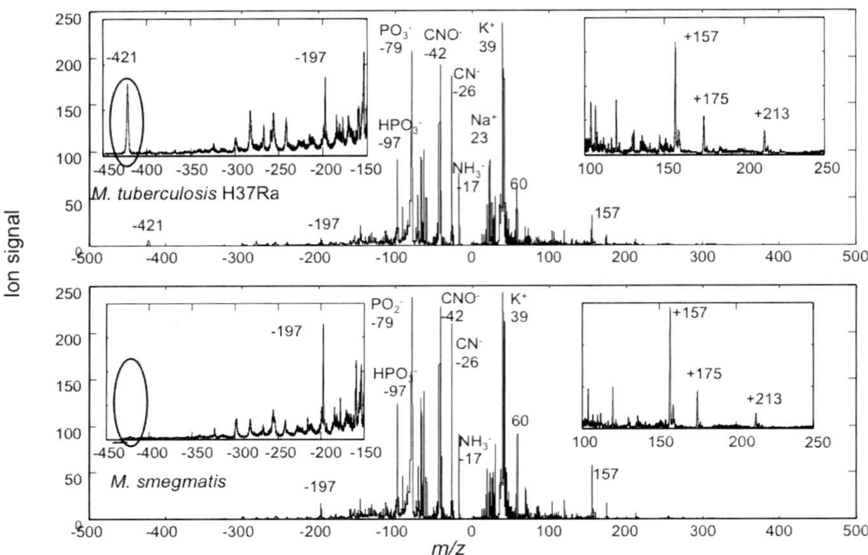

Figure 11. SPAMS Spectra for Mycobacterium tuberculosis H37Ra (top) and M. smegmatis (bottom). Averages of ~1000 single-particle spectra average are shown for improved visibility of discussed marker peaks. Peaks at m/z +157, +175 and 213 can be used to discriminate the two mycobacterial species shown here from B. atrophaeus and B. cereus vegetative cells (not shown in this figure). The peak at m/z = -421 is prominent in the majority (~66%) of single-particle spectra M. tuberculosis H37Ra and discriminates it from M. smegmatis.

that permitted direct counts of cells. These wind tunnel experiments had the goal of evaluating the SPAMS detection thresholds and aerosol concentration quantification capabilities. In the relatively background-free environment of the wind tunnel, SPAMS was able to sample and detect *M. tuberculosis* H37Ra at airborne concentrations of >1 *M. tuberculosis* H37Ra-containing particles/liter of air in 20 min as determined by direct counts of filter cassette-sampled particles, and concentrations of >40 *M. tuberculosis* H37Ra CFU/liter of air in 1 min as determined by using viable Andersen six-stage samplers. (It should be noted, that an early generation SPAMS instrument with simple converging nozzle particle inlet was used for these experiments and that newer SPAMS instruments, equipped with aerodynamic lens based particle inlets, have much improved particle throughput and hence sensitivity.)

The second set of experiments had the goal to evaluate rapid detection and identification of *Mycobacteria tuberculosis* in respiratory effluents with SPAMS (*15*). Again, *M. tuberculosis* H37Ra (TBa) was used as a surrogate for the virulent *M. tuberculosis* and *M. smegmatis* (MSm) was used as a bacterial near neighbor confounder. The mycobacteria were aerosolized from a matrix simulating respiratory effluents including either exhaled breath condensate (EBC) or bovine lung surfactant (LS) to produce conglomerate aerosol particles (single particles containing both respiratory effluents and bacteria). For reference, aerosols were also generated from the mycobacteria in aqueous solutions, from the respiratory

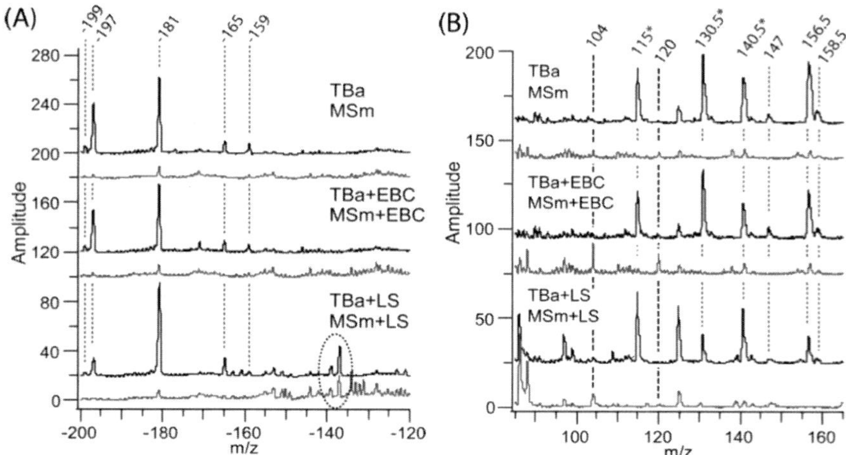

Figure 12. Comparison of ions detected from conglomerate particles containing M. tuberculosis H37Ra (TBa) and M. Smegmatis (MSm) over m/z ranges -200 to -120 (A) and 85 to 160 (B). Spectra from six types of particles are shown, TBa, TBa+EBC (exhaled breath condensate), TBa+LS (lung surfactant), MSm, MSm+EBC, and MSm+LS. Spectra are offset and labels rounded to the nearest half digit for clarity. Guide lines highlight TBa persistent peaks (···) and MSm persistent peaks (---) with m/z values labeled above the spectra. Asterisks represent peaks that are also observed in spectra from media. The dashed circle identifies LS associated ions detected in LS-containing conglomerate particles. Regions shown illustrate the largest amplitude TBa persistent peaks observed: -197, -181, 115, and 131* (from ref. (15)).*

matrices alone, and from bacterial growth media alone. The resulting aerosols were introduced into a SPAMS system and analyzed. SPAMS mass spectra of single conglomerate particles were shown to exhibit ions associated with both respiratory effluents and mycobacteria. It was also shown that several distinct mass spectral features distinguishing *M. tuberculosis* H37Ra from *M. smegmatis* in pure form also persist when conglomerate particles are analyzed. Figure 12 shows some of the results from this work.

In this work, a SPAMS alarm algorithm based on a modified pattern matching scheme was developed that blanked out known mass spectral peaks from media in order to base the bacterial discrimination on peaks from the two mycobacterial species alone (and not potential differences in growth media) With this algorithm SPAMS was able to distinguish *M. tuberculosis* H37Ra-containing particles and *M. smegmatis* from each other for >70% of the "pure" test particles and for >50% of the conglomerate particles containing not only bacteria but respiratory matrix (EBC or LS). Even though this discrimination is not perfect it is sufficient to enable a high probability of detection and a low false alarm rate if an adequate number of such particles are present and analyzed. For example, given a sample of 100 conglomerate particles and an alarm threshold of >23 particles, *M. tuberculosis* H37Ra can be identified in a background of *M. smegmatis* with a false positive rate of less than 10^{-5} according to binomial statistics.

The results from these two sets of experiments indicate the potential application of SPAMS for the direct detection of *M. tuberculosis*-containing aerosols generated by an infectious individual and its potential usefulness in rapid, reagentless tuberculosis screening. However, the current results from the proof-of-concept experiments described here do not directly translate to detection of tuberculosis in humans. Many challenges still exist in realizing rapid tuberculosis detection with a SPAMS-type system in clinical settings. These include investigating mass spectral patterns from actual virulent *M. tuberculosis* embedded in sputum from actual patients and the potentially large patient-to-patient variations in number of tuberculosis bacilli found in breath or sputum, as these natural variations may greatly affect detection capabilities.

Field Applications of SPAMS

Over the last several years, various versions of SPAMS systems were evaluated in numerous field tests to assess the systems' performance in operating autonomously for extended periods of time and in real-world aerosol backgrounds.

One of the early field tests was conducted at the San Francisco International Airport and had the goal to study aerosol backgrounds and potential sources of false alarms for other types of biosensors that were based on fluorescent particle detection. During this test, an early version of the SPAMS system containing a single fluorescence stage (with 266 nm excitation) operated in an air handler for one of the airport terminals for nearly 7 weeks in a semi-autonomous fashion (*56*, *57*). It should be noted that the fluorescence stage was not operated in pre-screening mode in this field test. Although particles were probed by the fluorescence stage for potential fluorescence the mass spectrometer was triggered by the tracking stage only and, thus, both fluorescent and non-fluorescent particles were analyzed by mass spectrometry. While the system was monitored from a remote site and had the capability to be controlled remotely, it ran autonomously for the entire period. During this ~7-week period the system operated normally >90% of the total time. Other than the weekly 1-hour maintenance (for system cleaning) there was only one significant down time (~3.5 days) due to some debris stuck in the particle tracking stage rendering the stage nearly inoperable. Figure 13 shows some of the results from this field test. Figure 13 (a) shows the aerosol particle concentration (in particles/cm^3) in the size range of 0.5-20 μm as measured by a TSI aerodynamic particle sizer (TSI Model 3321 APS) that was used as one of the reference instruments for parts of the 7-week period. Figure 13 (b) shows the number of particle mass spectra measured by SPAMS per 1-minute interval (which should be roughly proportional to the total particle count). Figure 13 (c) shows the number of fluorescent particles (i.e., being deemed "fluorescent" by the SPAMS fluorescent stage) among those particles shown in Figure 13 (b). (Note: the total number of particles tracked, sized and measured by the SPAMS fluorescence stage was much larger. Shown here in Figure 13 (b) and (c) are only those particles from which also SPAMS mass spectra were obtained.) The boxes in Figure 13 (b) and (c) indicate the ~3.5 day region where the SPAMS system operated with much decreased sensitivity due to the problems mentioned above. (This problem was noticed from the remote control location almost

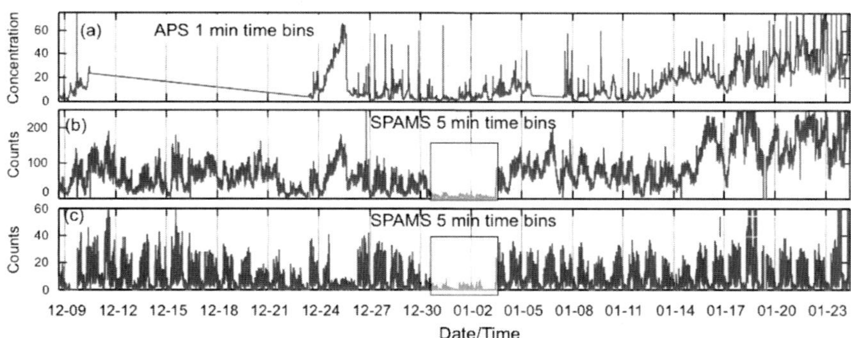

Figure 13. Results from a SPAMS field test and aerosol background study at the San Francisco International Airport. (a) aerosol particle concentration (in particles/cm³) in the size range of 0.5-20 um as measured by a TSI aerodynamic particle sizer (APS Model 3321) that was used as one of the reference instruments for parts of the 7-week period. (b) number of particle mass spectra measured by SPAMS per 1-minute interval (which should be roughly proportional to the total particle count). (c) number of fluorescent particles (i.e., being deemed "fluorescent" by the SPAMS fluorescent stage) among those particles shown in Figure 13 (b).

instantaneously, but could not be fixed right away because of access restrictions during the New Year's holiday period.)

From Figure 13 (a) and (b) it can be seen that there are large variations in total particle concentration across the whole 7-week period. Overall, there is a good correlation between the APS and SPAMS particle counts indicating that the SPAMS system operated at roughly constant sensitivity. Some of the particle concentration spikes appear more distinct in the APS particle counts. Further investigation revealed that those were caused primarily by relatively small particles (0.5 – 1.0 μm) that SPAMS is less sensitive to. As can be seen in Figure 13 (c), the SPAMS fluorescence stage revealed a pronounced diurnal cycle for fluorescent particles – a much more regular cycle than the total particle concentration. The majority of those fluorescent particles were attributed to aerosols produced in kitchen operations in an airport restaurant below the air handler. (Note that the restaurant closed early on 12/24 and was closed 12/25, which seems to be reflected in the SPAMS fluorescent particle counts.)

During this field test, nearly 1 million single aerosol particle spectra were collected. Interestingly, less than 1000 of these spectra were identified by SPAMS pattern recognition as spectra from *Bacillus* spores or vegetative bacteria. Further detailed analysis of those spectra revealed that they differed noticeably from the typical spore or vegetative cell spectra, respectively, and may have been misidentifications. The number of actual spores or vegetative cells in the ambient air during this measurement may have been even lower than what one would calculate from the number of spectra identified as spores or cells. Similar results were obtained in subsequent studies involving newer SPAMS systems and a much larger number of single-particle spectra collected over comparable time scales. These results indicate that the concentration of such bacteria at this location is relatively low and that the relatively frequent false alarms of other, purely fluorescence-based biosensors under evaluation during those field tests may have other sources. Further analysis of SPAMS data from the field and lab measurements on standards attributed some of the false alarms of other fluorescence-based sensors to other types (non-biological) of fluorescent aerosols.

Newer versions of the SPAMS system have demonstrated truly autonomous operation for extended periods of time that included real-time data analysis and running the alarm algorithm over the data and automatically adjusting alarm thresholds if the measured backgrounds warranted that step. While most of those SPAMS field tests were targeted at detection of biological materials, SPAMS has also been explored for the detection of other threat materials including chemical agents, explosives, radiological materials and drugs. More recently, we demonstrated that SPAMS should be able to detect and identify a broad range of threat material types using the same system setting and the same threat library and same set of alarm algorithms (*22*). While these results await confirmation in extensive field tests, they already indicate that SPAMS has the potential to be a near "universal detector".

Conclusions

Single-particle aerosol mass spectrometry as developed at LLNL over the last decade, is a powerful technology that reagentlessly analyzes individual aerosol particles, one-by-one at high rate. The direct UV laser desorption/ ionization technique used in SPAMS to obtain mass spectral signatures from single cells and aerosol particles does not produce the higher-mass signatures common in proteomics applications of MALDI-MS. Nevertheless, the smaller mass (<500 Da) chemical and metabolic signatures obtained with SPAMS in two polarities simultaneously provide a rich mass spectral fingerprint that allows differentiation of numerous types of biological aerosols, background aerosols and common interferents. Furthermore, SPAMS can even provide species discrimination between certain microorganisms.

One of the primary benefits of single particle analysis versus bulk analysis is that mixed aerosols or samples can be studied effectively. Whereas a single spore or cell might be lost in background clutter during bulk analysis, that spore or cell could be analyzed essentially free of background in a single particle instrument. They key, of course, is that the single particle instrument must be fast enough to analyze enough particles to actually find a particle of interest in a reasonable amount of time. This is why single particle detection and identification in combination with high-rate sampling, tracking and florescence prescreening is so useful.

Recently, it has been demonstrated that multiple threat particle types including biological particles, chemical weapons simulants, explosives particles, drugs, and metal particles (simulating radioactive materials) can be detected and automatically identified with a SPAMS system with the same system settings rendering SPAMS a near "universal detector." Moreover, measurements on conglomerate samples (where individual aerosol particles are mixtures or conglomerates) show evidence that SPAMS can also identify agents in preparations that the system was not specifically trained for as long as specific identifying mass spectral peaks can be observed.

SPAMS could also be useful as a quantitative tool or reference instrument. As we have shown in previous work that could not be described in more detail here, SPAMS can determine the concentrations of an individual particle type within mixtures of various other types of particles with vastly different individual concentrations. This is because the efficiency of each stage in a SPAMS system (and hence the efficiency of the entire system) as function of aerosol type and particle size can be determined in controlled laboratory measurements. Together with a comprehensive SPAMS performance model that we have developed these efficiencies can be used to calculate actual environmental concentrations of a particular particle type from the measured rate at which the corresponding mass spectra are acquired.

While SPAMS was originally developed for biodefense, it should not only be viewed as a counter-terrorism and biodefense tool. SPAMS can also be a basic research tool that is used to investigate aerosol particle compositions in the environment, particle-to-particle variation and aerosol aging. SPAMS also can provide means to analyze small quantities of samples introduced in micro-particle or droplet form and analyzed by mass spectrometry in the absence of a substrate. Lastly, SPAMS may also be a useful tool for studying fundamental gas-phase chemistry and competing gas phase acid-base reactions.

As the commercial development of new SPAMS systems and other single-particle aerosol mass spectrometers continues to progress, system costs should become comparable to or even lower than other routine mass spectrometry systems currently used in analytical labs. Single-particle aerosol mass spectrometry could, therefore, provide an attractive complement to other analytical equipment used in user-facilities for biological and chemical research.

Acknowledgments

Early proof-of-concept work for biodefense and biomedical applications of SPAMS were supported by the LLNL LDRD (Laboratory Directed Research and Development) Program under projects 02-ERD-002 and 05-ERD-053. The SPAMS system development was supported by the Technical Support Working Group (TSWG) and the Defense Advanced Research Project Agency (DARPA), both of the U.S. Department of Defense. The airport background study described here was supported by the U.S. Department of Homeland Security. Technical and logistical assistance during this study was also provided by collaborators from Sandia National Laboratory and personnel at the San Francisco airport and is greatly appreciated.

This work was performed under the auspices of the U. S. Department of Energy by Lawrence Livermore National Laboratory in part under Contract No. W-7405-Eng-48 and in part under DE-AC52-07NA27344.

References

1. Walt, D. R.; Franz, D. R. Biological Warfare Detection. *Anal. Chem.* **2001**, *72*, 738A–746A.
2. Fenselau, C.; Demirev, P. A. Characterization of intact micro-organisms by MALDI mass spectrometry. *Mass Spectrom. Rev.* **2001**, *20*, 157–171.
3. Demirev, P. A.; Fenselau, C. Mass spectrometry for rapid characterization of microorganisms. *Annu. Rev. Anal. Chem.* **2008**, *1*, 71–93.
4. Prather, K. A.; Nordmeyer, T.; Salt, K. Real-Time Characterization of Individual Aerosol-Particles Using Time-Of-Flight Mass-Spectrometry. *Anal. Chem.* **1994**, *66*, 1403–1407.
5. Hinz, K.-P.; Kaufmann, R.; Spengler, B. Laser Induced Mass Analysis of Single Particles in the Airborne State (LAMPAS). *Anal. Chem.* **1994**, *66*, 2071–2076.
6. Johnston, M. V.; Wexler, A. S. Mass Spectrometry of Individual Aerosol Particles. *Anal. Chem.* **1995**, *67*, 721A.
7. Noble, C. A.; Prather, K. A. Real-time single particle mass spectrometry: A historical review of a quarter century of the chemical analysis of aerosols. *Mass Spectrom. Rev.* **2000**, *19*, 248–274.
8. Johnston, M. V. Sampling and analysis of individual particles by aerosol mass spectrometry. *J. Mass Spectrom.* **2000**, *35*, 585–595.
9. Murphy, D. M.; Cziczo, D. J.; Froyd, K. D.; Hudson, P. K.; Matthew, B. M.; Middlebrook, A. M.; Peltier, R. E.; Sullivan, A.; Thomson, D. S.; Weber, R. J. Single-particle mass spectrometry of tropospheric aerosol particles. *J. Geophys. Res.* **2006**, *111*, D23S32.
10. Canagaratna, M. R.; Jayne, J. T.; Jiménez, J. L.; Allan, J. D.; Alfarra, M. R.; Zhang, Q.; Onasch, T. B.; Drewnick, F.; Coe, H.; Middlebrook, A.; Delia, A.; Williams, L. R.; Trimborn, A. M.; Northway, M. J.; DeCarlo, P. F.; Kolb, C. E.; Davidovits, P.; Worsnop, D. R. Chemical and Microphysical Characterization of Ambient Aerosols with the Aerodyne Aerosol Mass Spectrometer. *Mass Spectrom. Rev.* **2007**, *26*, 185–222.
11. Russell, S. C. Microorganism characterization by single particle mass spectrometry. *Mass Spectrom. Rev.* **2009**, *28*, 376–387.
12. Fergenson, D. P.; Pitesky, M. E.; Tobias, H. J.; Steele, P. T.; Czerwieniec, G. A.; Russell, S. C.; Lebrilla, C. B.; Horn, J. M.; Coffee, K. R.; Srivastava, A.; Pillai, S. P.; Shih, M. P.; Hall, H. L.; Ramponi, A. J.; Chang, J. T.; Langlois, R. G.; Estacio, P. L.; Hadley, R. T.; Frank, M.; Gard, E. E. Reagentless Detection and Classification of Individual Bioaerosol Particles in Seconds. *Anal. Chem.* **2004**, *76*, 373–378.
13. Tobias, H. J.; Pitesky, M. E.; Fergenson, D. P.; Steele, P. T.; Horn, J. M.; Frank, M.; Gard, E. E. Following the metabolic and morphological changes of individual Bacillus atrophaeus cells during the sporulation process using Bioaerosol Mass Spectrometry. *J. Microbiol. Methods* **2006**, *67*, 56–63.
14. Tobias, H. J.; Schafer, M. P.; Pitesky, M. E.; Fergenson, D. P.; Horn, J.; Frank, M.; Gard, E. E. Bioaerosol Mass Spectrometry (BAMS) for the Rapid Detection of Individual Airborne Mycobacterium tuberculosis H37Ra Particles. *Appl. Environ. Microbiol.* **2005**, *71*, 6086–6095.

15. Adams, K. L.; Steele, P. T.; Bogan, M. J.; Sadler, N. M.; Martin, S. I.; Martin, A. N.; Frank, M. Reagentless Detection of Mycobacteria tuberculosis H37Ra in Respiratory Effluents in Minutes. *Anal. Chem.* **2008**, *80*, 5350–5357.
16. Ferge, T.; Karg, E.; Schröppel, A.; Coffee, K. R.; Tobias, H. J.; Frank, M.; Gard, E. E.; Zimmermann, R. Fast determination of the relative elemental and organic carbon content of aerosol samples by single-particle aerosol time-of-flight mass spectrometry. *Environ. Sci. Technol.* **2006**, *40*, 3327–3335.
17. Martin, A. N.; Farquar, G. R.; Gard, E. E.; Frank, M.; Fergenson, D. P. Identification of High Explosives Using Single Particle Aerosol Mass Spectrometry. *Anal. Chem.* **2007**, *79*, 1918–1925.
18. Martin, A. N.; Farquar, G. R.; Gard, E. E.; Frank, M.; Fergenson, D. P. Single Particle Aerosol Mass Spectrometry for the Detection and Identification of Chemical Warfare Agent Simulants. *Anal. Chem.* **2007**, *79*, 6368–6375.
19. Barker, Z.; Venkatachalam, V.; Martin, A. N.; Farquar, G. R.; Frank, M. Detecting Trace Pesticides in Real Time Using Single Particle Aerosol Mass Spectrometry. *Anal. Chim. Acta* **2010**, *661*, 188–194.
20. Martin, A. N.; Farquar, G. R.; Jones, A. D.; Frank, M. The Non-Destructive Identification of Solid Over-the-Counter Medications using Single Particle Aerosol Mass Spectrometry. *Rapid Commun. Mass Spectrom.* **2007**, *21*, 3561–3568.
21. Martin, A. N.; Farquar, G. R.; Steele, P. T.; Jones, A. D.; Frank, M. The use of single particle aerosol mass spectrometry for the automated non-destructive identification of drugs in complex samples. *Anal. Chem.* **2009**, *81*, 9336–9342.
22. Steele, P. T.; Farquar, G. R.; Martin, A. N.; Coffee, K. R.; Riot, V. J.; Martin, S.; Fergenson, D. P.; Gard, E. E.; Frank, M. Autonomous, Broad-Spectrum Detection of Hazardous Aerosols in Seconds. *Anal. Chem.* **2008**, *80*, 4583–4589.
23. Gard, E. E.; Coffee, K. R.; Frank, M.; Tobias, H. J.; Fergenson, D. P.; Madden, N.; Riot, V. J.; Steele, P. T.; Woods, B. W. Improved real-time detection method and system for identifying individual aerosol particles. US Patent 7,260,483, 2007.
24. Schreiner, J.; Schild, U.; Voigt, C.; Mauersberger, K. Focusing of aerosols into a particle beam at pressures from 10 to 150 Torr. *Aerosol Sci. Technol.* **1999**, *31*, 373–382.
25. Riot, V. J. LLNL, 2004, unpublished.
26. Wang, X.; Kruis, F. E.; McMurry, P. H. Aerodynamic focusing of nanoparticles: I. Guidelines for designing aerodynamic lenses for nanoparticles. *Aerosol Sci. Technol.* **2005**, *39*, 611–623.
27. Gard, E. E.; Riot, V. J.; Coffee, K. R.; Woods, B. W.; Tobias, H. J.; Birch, J.; Weisgraber, T. Pressure-flow reducer for aerosol focusing devices. US Patent 7,361,891, 2008.
28. Farquar, G. R.; Steele, P. T.; McJimpsey, E. L.; Lebrilla, C. B.; Tobias, H. J.; Gard, E. E.; Frank, M.; Coffee, K. R.; Riot, V. J.; Fergenson, D. P. Supramicrometer particle shadowgraph imaging in the ionization region of a single particle aerosol mass spectrometer. *J. Aerosol Sci.* **2008**, *39*, 10–18.

29. Noble, C. A.; Prather, K. A. Real-Time Measurement of Correlated Size and Compostion Profiles of Individual Atmospheric Aerosol Particles. *Environ. Sci. Technol.* **1996**, *30*, 2667–2680.
30. Hill, S. C.; Pinnick, R. G.; Niles, S.; Pan, Y.-L.; Holler, S.; Chang, R. K.; Bottiger, J.; Chen, B. T.; Orr, C.-S.; Feather, G. Real-time measurement of fluorescence spectra from single airborne biological particles. *Field Anal. Chem. Technol.* **1999**, *3*, 221–239.
31. Stowers, M. A.; van Wuijckhuijse, A. L.; Marijnissen, J. C. M.; Kientz, Ch. E; Chiach, T. Fluorescence preselection of bioaerosol for single-particle mass spectrometry. *Appl. Optics* **2006**, *45*, 8531–8536.
32. Steele, P. T.; Tobias, H. J.; Fergenson, D. P.; Pitesky, M. E.; Horn, J. M.; Czerwieniec, G. A.; Russell, S. C.; Lebrilla, C. B.; Gard, E. E.; Frank, M. Laser Power Dependence of Mass Signatures from Individual Bacterial Spores in Bioaerosol Mass Spectrometry. *Anal. Chem.* **2003**, *75*, 5480–5487.
33. Steele, P. T.; Srivastava, A.; Pitesky, M. E.; Fergenson, D. P.; Tobias, H. J.; Gard, E. E.; Frank, M. Desorption/Ionization Fluence Thresholds and Improved Mass Spectral Consistency Measured Using a Flattop Laser Profile in the Bioaerosol Mass Spectrometry of Single Bacillus Endospores. *Anal. Chem.* **2005**, *77*, 7448–7454.
34. Czerwieniec, G. A.; Russell, S. C.; Coffee, K. R.; Riot, V. J.; Steele, P. T.; Frank, M.; Gard, E. E.; Lebrilla, C. B. Improved Sensitivity and Mass Range in Time-of-Flight Bio-Aerosol Mass Spectrometry Utilizing an Electrostatic Ion Guide. *J. Am. Soc. Mass Spectrom.* **2005**, *16*, 1866–1875.
35. Stowers, M. A.; van Wuijckhuijse, A. L.; Marijnissen, J. C. M.; Scarlett, B.; van Baar, B. L. M.; Kientz, Ch. E Application of matrix-assisted laser desorption/ionization to on-line aerosol time-of-flight mass spectrometry. *Rapid Commun. Mass Spectrom.* **2000**, *14*, 829–833.
36. Jackson, S. N.; Mishra, S.; Murray, K. K On-line laser desorption/ ionization mass spectrometry of matrix-coated aerosols. *Rapid Commun. Mass Spectrom.* **2004**, *18*, 2041–2045.
37. Kim, J. K.; Jackson, S. N.; Murray, K. K. Matrix-assisted laser desorption/ionization mass spectrometry of collected bioaerosol particles. *Rapid Commun. Mass Spectrom.* **2005**, *19*, 1725–1729.
38. Harris, W. A.; Reilly, P. T. A.; Whitten, W. B. MALDI of individual biomolecule-containing airborne particles in an ion trap mass spectrometer. *Anal. Chem.* **2005**, *77*, 4042–4050.
39. Kleefsman, W. A.; Stowers, M. A.; Verheijen, P. J. T.; Marijnissen, J. C. M. Single Particle Mass Spectrometry-Bioaerosol Analysis by MALDI MS. *KONA Powder and Particle Journal* **2008**, *26*, 205–214.
40. Russell, S. C.; Czerwieniec, G.; Lebrilla, C.; Steele, P. T.; Riot, V. J.; Coffee, K. R.; Frank, M.; Gard, E. E. Achieving High Detection Sensitivity (14 zmole) of Biomolecular Ions in Bioaerosol Mass Spectrometry. *Anal. Chem.* **2005**, *77*, 4734–4741.
41. Bogan, M. J.; Patton, E.; Srivastava, A.; Martin, S.; Fergenson, D. P.; Steele, P. T.; Tobias, H. J.; Gard, E. E.; Frank, M. Online aerosol mass

spectrometry of single micrometer-sized particles containing poly(ethylene glycol). *Rapid Commun. Mass Spectrom.* **2007**, *21*, 1214–1220.

42. McJimpsey, E. L.; Jackson, W. M.; Lebrilla, C. B.; Tobias, H. J.; Bogan, M. J.; Gard, E. E.; Frank, M.; Steele, P. T. Parameters Contributing to Efficient Ion Generation in Aerosol MALDI Mass Spectrometry. *J. Am. Soc. Mass Spectrom.* **2008**, *19*, 315–324.
43. Wade, E. E.; Farquar, G. R.; Steele, P. T.; McJimpsey, E. L.; Lebrilla, C. B.; Fergenson, D. P. Wavelength and size dependence in single particle laser aerosol mass spectra. *Aerosol Sci.* **2008**, *39*, 657–666.
44. Riot, V. J.; Coffee, K. R.; Gard, E. E., Fergenson, D. P.; Ramani, S.; Steele, P. T. DSP-based dual-polatity mass spectrum pattern recognition for bio-detection.*IEEE Proceedings 4th IEEE Workshop on Sensor Array and Multichannel Processing*, 2006, 1-4244-0309-X/06/98-101.
45. Westphal, A. J.; Price, P. B.; Leighton, T. J.; Wheeler, K. E. Kinetics of size changes of individual *Bacillus thuringiensis* spores in response to changes in relative humidity. *Proc. Natl. Acad. Sci. U.S.A.* **2003**, *100*, 3461–3466.
46. Czerwieniec, G. A.; Russell, S. C.; Tobias, H. J.; Pitesky, M. E.; Fergenson, D. P.; Steele, P. T.; Srivastava, A.; Horn, J. M.; Frank, M.; Gard, E. E.; Lebrilla, C. B. Stable Isotope Labeling of Entire Bacillus atrophaeus Spores and Vegetative Cells Using Bioaerosol Mass Spectrometry. *Anal. Chem.* **2005**, *77*, 1081–1087.
47. Srivastava, A.; Pitesky, M. E.; Steele, P. T.; Tobias, H. J.; Fergenson, D. P.; Horn, J. M.; Russell, S. C.; Czerwieniec, G. A.; Lebrilla, C. B.; Gard, E. E.; Frank, M. Comprehensive Assignment of Mass Spectral Signatures from Individual *Bacillus atrophaeus* Spores in Matrix-Free Laser Desorption/Ionization Bioaerosol Mass Spectrometry. *Anal. Chem.* **2005**, *77*, 3315–3323.
48. Russell, S. C.; Czerwieniec, G. A.; Lebrilla, C. B.; Tobias, H. J.; Fergenson, D. P.; Steele, P. T.; Pitesky, M. E.; Horn, J. M.; Frank, M.; Gard, E. E. Toward Understanding the Ionization of Biomarkers from Micrometer Particles by Bioaerosol Mass Spectrometry. *J. Am. Soc. Mass Spectrom.* **2004**, *15*, 900–909.
49. Wissner-Gross, Z. D.; Farquar, G. R.; Steele, P. T.; Martin, A. N.; Bogan, M. J.; Wade, E. A.; Tobias, H. J.; Fergenson, D. P.; Frank, M. Signal Variation in Single Particle Aerosol Mass Spectrometry. Unpublished, 2010.
50. Ryzhov, V.; Hathout, Y.; Fenselau, C. Rapid Characterization of Spores of *Bacillus cereus* Group Bacteria by Matrix-Assisted Laser Desorption-Ionization Time-of-Flight Mass Spectrometry. *Appl. Environ. Microbiol.* **2000**, *66*, 3828–3834.
51. Ong, S.-E.; Mann, M. Mass spectrometry–based proteomics turns quantitative. *Nat. Chem. Biol.* **2005**, *1*, 252–262.
52. Meetani, M. A.; Shin, Y.-S.; Zhang, S.; Mayer, R.; Basile, F. Desorption electrospray ionization mass spectrometry of intact bacteria. *J. Mass Spectrom.* **2007**, *42*, 1186–1193.
53. Srivastava, A.; Martin, S. I.; Steele, P. T.; Tobias, H. J.; Gard, E. E.; Frank, M. Osmoadaptation in *Bacillus* Species Studied using Single Particle Aerosol Mass Spectrometry. Unpublished, 2010.

54. Taborsky, V. Small-scale processing of microbial pesticides. *FAO Agricultural Services Bulletin* **1992**, 96 (Food and Agriculture Organization of the United Nations, Rome, 1992, see http://www.fao.org/docrep/t0533e/t0533e00.htm).
55. Rebotier, T. P.; Prather, K. A. Aerosol time-of-flight mass spectrometry data analysis: A benchmark of clustering algorithms. *Anal. Chim. Acta* **2007**, *585*, 38–54 and references therein.
56. Steele, P. T.; McJimpsey, E. L.; Coffee, K. R.; Fergenson, D. P.; Riot, V. J.; Tobias, H. J.; Woods, B. W.; Gard, E. E.; Frank, M. Characterization of Ambient Aerosols at the San Francisco International Airport Using BioAerosol Mass Spectrometry. *Proc. SPIE* **2006**, 6218 (*Chemical and Biological Sensing VII*, 62180A).
57. McJimpsey, E. L.; Steele, P. T.; Coffee, K. R.; Fergenson, D. P.; Riot, V. J.; Woods, B. W.; Gard, E. E.; Frank, M.; Tobias, H. J.; Lebrilla, C. Detection of Biological Particles in Ambient Air Using BioAerosol Mass Spectrometry. *Proc. SPIE* **2006**, 6218 (*Chemical and Biological Sensing VII*, 62180B).

Chapter 11

Rapid Detection and Identification of Aerosolized Biological Materials, Toxins, and Microorganisms by an AP-MALDI-MS-Based System

Berk Oktem,* Appavu K. Sundaram, Jane Razumovskaya, Seshu K. Gudlavalleti, Thomas D. Saul, and Vladimir M. Doroshenko

Science & Engineering Services Inc., Columbia, MD
***oktem@apmaldi.com**

We previously reported development of a proteomics and AP-MALDI-MS/MS based system for biological detection and identification, offering fast analysis in minutes. In this work, a direct aerosol sampling module is reported which is interfaced with a small sample preparation module to reduce the overall size and power consumption. Bioaerosol samples are directly collected on a conveyor surface. On the same surface, sample processing is conducted, which include extraction of target-specific proteins using selective solubilization methods and *in situ* proteolysis. The microorganisms or toxins are identified using an AP-MALDI ion trap MS and MS/MS typing method. Identification for close relatives BA and BC with the MS/MS typing method is described.

Introduction

Intentional release of biological aerosols in large quantities is a major concern due to the silent nature of such discharge. Noticable effects are often observed too late to administer effective treatment. The scope of the release can only be realized upon monitoring hospital records that have an unusual increase of similar clinical symptoms. Pathogenic microorganisms such as *Bacillus anthracis*, which causes anthrax, *Variola* which is the virus that causes smallpox, or ricin which is a toxin can be released in aerosol phase as dry powder (*1–3*).

Other concerns include detection and identification of nerve gases, blistering chemicals and toxic industrial chemicals (TICs). Most nerve gases such as Sarin, VX and toxic industrial chemicals are small volatile or semivolatile molecules that can be released in gas phase or adsorbed on soil. These can be harmful, even lethal upon inhalation or skin contact. In some cases, these chemicals are incapacitating (*4*).

Mass spectrometry (MS) has been used extensively in both biological and chemical detection (*5*, *6*). Due to soft ionization and higher tolerance towards impurities matrix assisted laser desorption ionization time-of-flight (MALDI-TOF) MS has been used for rapid analysis in biological detection. Methods initially developed for analysis of spores by Fenselau and co-workers (*7*, *8*) were then generalized to include viruses and protein toxins using atmospheric pressure matrix assisted laser desorption ionization (AP-MALDI) MS. These systems can employ ion trap MS analyzers with MS/MS capability (*9*, *10*).

The multifaceted nature of biological aerosols and gas-phase small molecules, warrant very different approaches for detection and identification. Our effort is to create a single platform that can detect both chemical and biological material within minutes. We previously reported development of a proteomics and AP-MALDI-MS/MS based system for biological detection and identification offering fast analysis in minutes (*9*). This report will focus on two separate experiments performed with the AP-MALDI-MS based method. The first is aerosolized microorganism detection with automated sample collection and analysis. Secondly, differentiation of close relatives *Bacillus anthracis* (BA) and *Bacillus cereus* (BC) is reported. BA and BC were studied in liquid phase samples without use of the automated system to investigate different options of MS/MS analysis.

Methods Overview

Aerosol Sampling

The system described here uses a high flow aerosol collector (400 L/min). It is a two stage virtual impactor based concentrator which directs the output of the concentrator to the collection surface.

0.5-10 μm sized particles are sampled through a 8" long, 1.25" diameter stainless steel tube into the 1st stage of the virtual impactor (*11*). A schematic is shown in Figure 1. Particles that are smaller than the cut-off size of the impactor will follow the air streamlines of the major flow (roughly 360 L/min). Due to their inertia, larger particles are retained in the minor flow, which is roughly 40 L/min. The minor flow of the first stage is introduced as the input flow of the second stage, where they are further concentrated. The minor flow in the second stage (around 5 L/min) is split into two nozzles that transmit same concentration of particles, therefore generating two identical 2 mm wide sample spots on the collection surface.

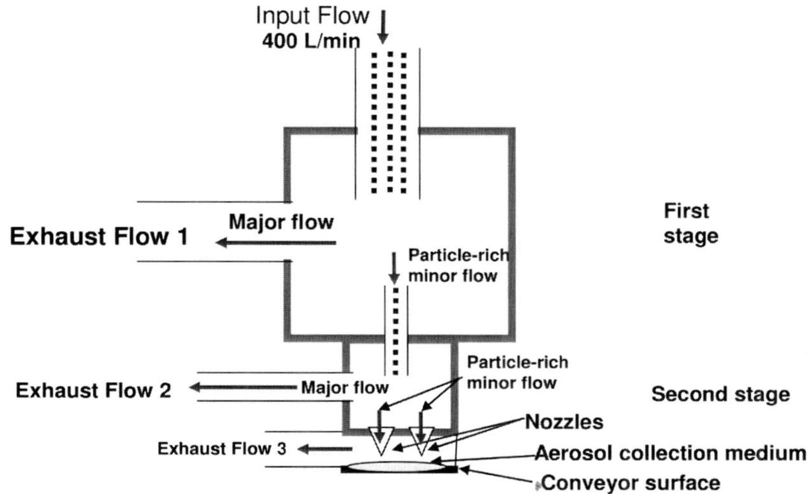

Figure 1. Aerosol sampling and collection setup.

Figure 1 also illustrates the sample collection surface. The incoming particle rich stream exiting the nozzles is directed at the collection surface where particles are expected to hit the surface at a speed around 20 m/s. Under these circumstances, particle bouncing is likely, as reported in cascade impactors (*12*). To reduce particle bouncing effect, we coat the surface with a thin aerosol collection medium (to be discussed below). This creates a viscous surface where particles are retained. Similar technique has been reported employing Apiezon-L grease and silicone oil for cascade impactors (*13*).

One of the challenges here was to ensure compatibility with AP-MALDI ionization. Similar to vacuum MALDI, sample preparation for AP-MALDI requires co-crystallization of matrix and analyte with a low salt content (*14*). Krytox (Dupont, Wilmington, DE) offered such an advantage. The coating on the surface did not adversely effect AP-MALDI analysis (data not shown). Typically 2-3 μL of suspension is deposited. Due to the volatility of the solvent, evaporation occurs in a matter of seconds leaving a thin layer of Krytox on the surface. An added benefit of Krytox is its hydrophobicity, which helps with the chemical processing of the samples. This will be described in the next section.

Two independent yet identical small diaphragm pumps provide the flow exiting the nozzle. This ensures that the flow going through each nozzle is balanced and have equal concentration of particles. Particles exiting the nozzle hit the collection surface, which is 3 mm away from the tip of the nozzle. The described aerosol collection method creates two 2 mm in diameter spots.

We have two ways to verify aerosol collection on the surface. One is MS analysis of the collected material. MS signal of the collected material can be compared to the MS signal obtained by depositing a solution containing a known amount of analyte. The second one is a small probe fluoresence detector, which will be described here.

The small probe fluoresence detector is equipped with a fiber-optic probe that can capture and quantify fluorescence from a surface (Opti-Sciences, Inc., Hudson, NH). Fluorescent polystyrene latex (PSL) spheres are used as sample aerosol (Thermo-Duke Scientific, Palo Alto, CA). When a solution containing these spheres are nebulized, monodisperse aerosol is created. Following collection, the fluorescent signal from the collected particles are compared with fluorescent signal of the control spot. Control spot is obtained with an identical surface containing a measured amount of fluorescent particles, e.g. 5 μL of the PSL solution is deposited with a pipettor to create 2 mm wide spot. Figure 2 shows the change in fluorescence signal as a function of aerosol deposition time, when the total particle concentration of the introduced PSL was 40 particles/cm^3. Efficiency of collection will be discussed further later in the text.

For testing, aerosol was generated with a bio-aerosol nebulizing generator, BANG (CH Technologies Inc., Westwood, NJ). The output was verified by an aerodynamic particle sizer, APS (Model 3321, TSI Inc., Shoreview, MN).

Sample Processing

Proteomics Based Analysis

Microorganisms and toxins are treated with multiple chemicals during sample processing. By AP-MALDI-MS, peptide ions and their MS/MS fragments are generated. A bioinformatics based data analysis isthen conducted. A commercial AP-MALDI source was used in this work (MassTech, Columbia, MD). This type of source can be used with most mass spectrometers with an atmospheric pressure ionization (API) inlet- such as ESI or APCI. In this work, an ion trap MS system was used for analysis (LCQ Deca XP, Thermo, Waltham, MA).

Figure 3 illustrates how the sample is processed. On-probe chemical processing takes place on a hydrophobic surface. Several extraction reagents enable targeting different classes of biological material. 10 % TFA treatment of spores enables observation of small acid-soluble proteins (SASPs), which was named 'spore protocol'. 50 % NH_4OH treatment of viruses enable observation of capsid proteins. If no reagent is used, protein toxins are detected. If spores are targeted but no acid or trypsin is added, we named this as 'whole spore protocol'. Finally, with 1% TFA in 50% acetonitrile, all of the above can be observed, therefore named 'unified broadband protocol'. These extracted proteins are then digested by trypsin and washed with water. Finally, matrix is added. The peptide fragments are then observed by AP-MALDI MS/MS.

Automation of Sample Preparation

Translating the lab-bench operation to a viable automation mechanism required additional strategies. In our earlier description, we presented a fully automated system (*9*). It featured two fast actuators for rapid sample processing which reduced travel from one end of the deck (containing samples, reagents, disposable pipette tips, and a sample processing block) to the other end -a distance

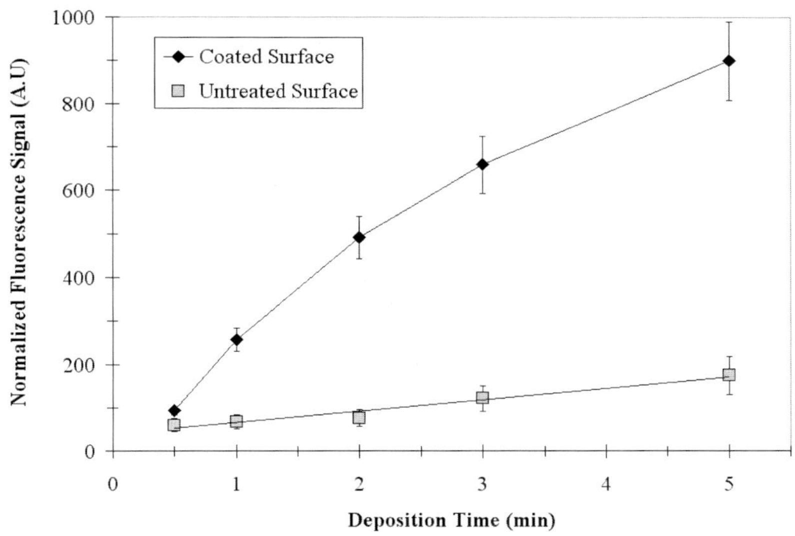

Figure 2. Fluorescence signal of collected 1.2 ±0.2 μm PSL aerosols as a function of deposition time on Krytox-coated and untreated surfaces.

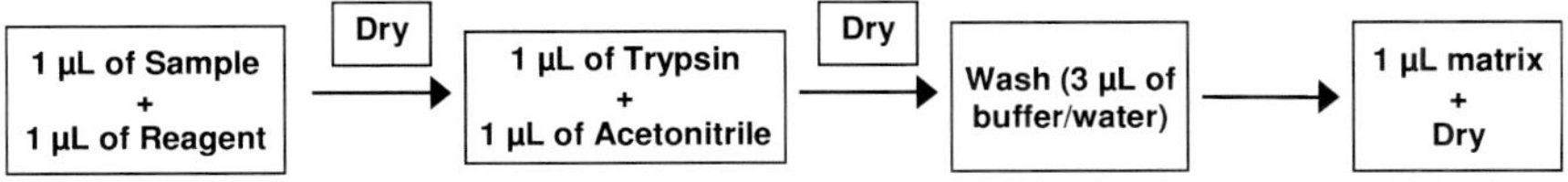

Agent specific extraction reagents:

Spores: 10% trifluoro acetic acid (TFA)
Virus: 50% ammonium hydroxide solution
Toxins: None
Unified Broadband: 1% trifluoro acetic acid (TFA) in 50% Acetonitrile

Figure 3. Schematic of biological sample processing.

of 80 cm- in 0.5 second. Samples and reagents with volumes of 1-10 μL are aspirated and dispensed by an automated pipettor lowered and raised with a third actuator (BioHIT, Helsinki, Finland). Using disposable tips (to eliminate possibility of cross-contamination) and an automated plate transfer between the processing block and the AP-MALDI stage are other features of the earlier system. Gold coated stainless steel plates are used for the processing surface.

However, that system had some disadvantages as well: the deck of the previous system had to contain several loads of pipette tips, which increased the area that needs to be traveled by the pipettor. This, along with the demand for faster processing, required fast actuators which consume more power. High power consumption, however, is not desired for field detection.

Due to the demand to achieve the smallest footprint of a new instrument with minimum consumables and waste, we pursued other alternatives. To eliminate travel required to reach both reagents and fresh pipette tips, a new contact-free dispenser is employed in our new system. Reagents are stored in separate containers, where each have a dedicated contact-free dispenser. This employs

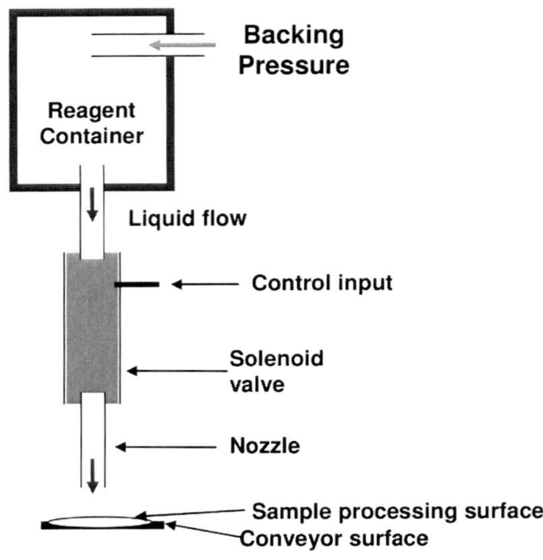

Figure 4. Schematic of the contact-free dispenser.

solenoid based valves, where liquid is driven by a constant backing pressure in the reagent container with values ranging 5-15 psi (Figure 4). Solenoid valves with 0.25 mm ID nozzles were kept at a height of 2-5 mm above the sample surface depending on the amount of solution dispensed. The solenoid valve opens for a period of 1-10 ms, which is controlled by a TTL pulse. For example, approximately 1 μL of water is dispensed with a backing pressure of 10 psi and when the valve is open for 4 ms.

To eliminate the need of fresh plates for sample processing, a conveying video tape is used as a collector surface. Automated sample processing is carried out on the same surface. Then, mass spectra are collected by AP-MALDI by irradiating the processed sample spot with no sample transfer. This also eliminates any sample transfer losses from one media to another. The tape material is conductive and it is able to withstand 3 kV high voltage, which is required for AP-MALDI operation. Additionally, the video tape is magnetic. A flat surface for sample processing and mass spectral analysis is ensured by placing magnets under the tape. During processing, the tape surface is held by latching solenoids and can be moved 5 mm in the x and 16 mm in the y direction, which is necessary to raster the sample spot coincidence to the laser beam. Extra tape length is provided before latching to accommodate this motion without tearing the tape.

The MS data acquisitions reported here used an ion trap mass analyzer, which has an AP-MALDI ion source as described earlier. Processed samples on the video tape were cut and attached to a regular APMALDI plate with double-sided sticky tape. A portable mass spectrometer is currently in development, which will be interfaced with the aerosolized sample processing module described here. This portable mass spectrometer development which features low power and mass consumption will be reported separately.

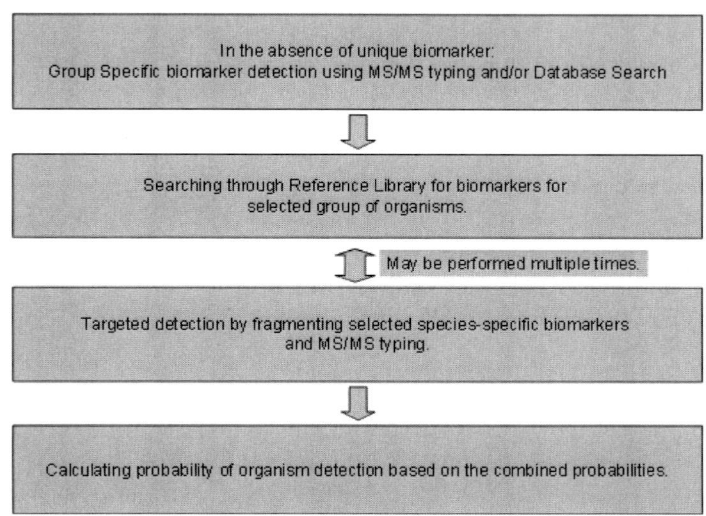

Figure 5. Overall strategy of organism detection and differentiating between close neighbors.

BA and BC analysis were performed without the use of automated analysis. Same protocols were used as described above. Liquid samples of BA and BC are obtained from ATCC (Manassas, VA).

Data Processing and Analysis

To ensure the quickest library search, while minimizing the number of false positive identifications, we developed a new methodology to process the data.

Some of the earliest approaches to biological identification by mass spectrometry involved the development of a library of MS (mass spectra) fingerprints for each biological material. The MS spectral fingerprint library approach has limited reliability. Peaks observed in the mass spectrum of the same biological material can vary significantly under different conditions, such as field dependent background or microorganism growth media.

Peptide matching by MASCOT or other MS/MS search engine were developed to provide more reliable identification. In this approach, the recorded MS/MS data is matched against theoretically derived peptide tandem MS spectra which come from available protein sequence databases. It further provides information on which proteins and organisms those peptides originate from. Thus, a unique match of peptide-to-protein and protein-to-organism can result in an unambiguous organism detection. MS/MS search engines are applied widely today in the field of proteomics for MS/MS data interpretation.

More recently, a new method for biological detection called *MS/MS typing* was developed (*9*). MS/MS typing is a bottom-up method for microorganism detection. The MS/MS signature of a species-specific biomarker, and its direct comparison with a reference MS/MS spectrum (usually obtained in a separate experiment) contained in the reference library is measured via spectral correlation techniques. Similar to MS spectral fingerprint method, it relies on MS/MS spectral libraries for detection. However, due to the insensitivity of MS/MS spectra to background and growth conditions, as well as its capacity to focus on wide range of specific biomarkers regardless of their nature (peptide, lipid, phospholipids, lipopeptides, etc.), MS/MS typing provides an extremely reliable and effective detection strategy. Furthermore, it encompasses probability assessment to assign confidence levels for each positive detection. Using MS/MS typing, we observed roughly an order of magnitude improvement in sensitivity for peptide detection in comparison to the traditional MS/MS search methods. Lipids, phospholipids and lipopeptides are likewise detected.

In our detection methodology, we apply MS/MS typing built into a biodetection algorithm. In real time, it coordinates interactive mass spectrometry data collection and analysis. It uses data dependent MS/MS data collection to minimize overall analysis time and maximize detection of multiple targets, while maintaining high sensitivity of detection and reducing time to generate red flags to indicate positive identifications. To improve the detection coverage, we apply a combination of MS/MS typing methodology (for detection of biomarkers recorded in the Reference Library) in conjunction with basic MS/MS search engine (MASCOT) set to search againset NCBInr protein sequence database (for detection of biomarker peptides not included in the reference library).

Here we show an example of using AP-MALDI MS/MS typing for rapid organism identification based on the detection of molecular biomarkers by comparing their MS/MS spectra to reference library spectra. This approach surpasses routine database search method for organism detection in terms of speed and sensitivity. It also allows including non-peptide and unknown (unsequenced or modified) peptide biomarkers in the detection. The latter advantage is invaluable for differentiating between close neighbors, such as *Bacillus anthracis* (BA) and *Bacillus cereus* (BC) spores, as in some cases there are no known unique biomarkers or they are observed in low-intensity peaks. The detection framework relies on tiered biomarker detection by MS/MS typing. In the absence of unique biomarker, the first tier of detection localizes the group of organisms represented by a non-unique (common to the group) biomarker. This is followed by a search for the presence/absence of biomarker not present in the proteome database, confirming the observed species. The process is outlined in the flowchart in Figure 5. The approach is not limited to any single chemical protocol, allowing for combinations of detection protocols, if deemed necessary. The probability of the organism detection is calculated based on the combined probabilities of detection/non-detection of each of the considered biomarkers. Performance of the methodology is evaluated by the detection of BA versus its close neighbors, BC ATCC 14579 and BC ATCC 10987.

Results and Discussion

Collection and Detection Efficiency of Aerosol Samples

Collection efficiency for particles of 0.5 μm to 2.5 μm size was measured with fluorescent latex spheres and BG spores (Figure 6). A $1 \cdot 10^8$ cfu/mL suspension

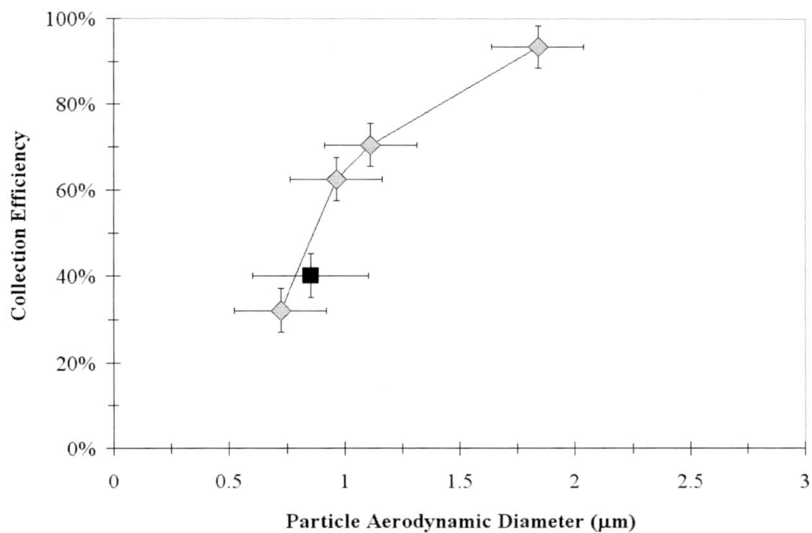

Figure 6. Particle collection efficiency onto Krytox-coated surface as a function of aerodynamic diameter. Points shown with diamonds are for PSL particles. Square point depicts the BG spore aerosol (d_{mean}~0.8μm).

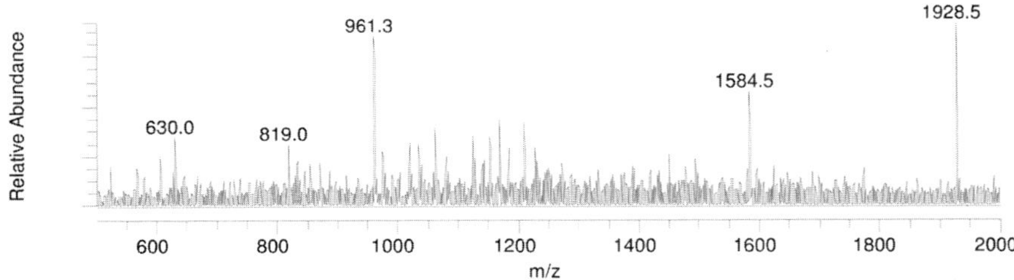

Figure 7. AP-MALDI-MS of aerosolized BG after extracting SASPs and digestion with trypsin.

(cfu: colony forming units) of BG spores is aerosolized by the nebulizer with a 5 psi backing pressure. We did not attempt to control the BG spores size. The mode diameter of the generated BG spores was around 0.8 μm measured by APS with a geometric standard deviation σ_g of 1.2 (data not shown). This is the typical size for individual BG spores considering the aerosol generation method and the concentration of BG used. With alternative aerosol generation methods or conditions, larger aerosol agglomerates with multiple BG spores per particle could be produced. The efficiency curtails with decreasing particle size as shown by other virtual impactors (*11*). We were able to account for those losses in the exhaust flow of the particles (data not shown).

Figure 7 shows the AP-MALDI MS spectrum of the collected BG particles, after proteomics sample processing as described earlier. The peaks at *m/z* 1585 and *m/z* 1929 are peptide ions from SASPs of BG as shown earlier (*9*). Based on

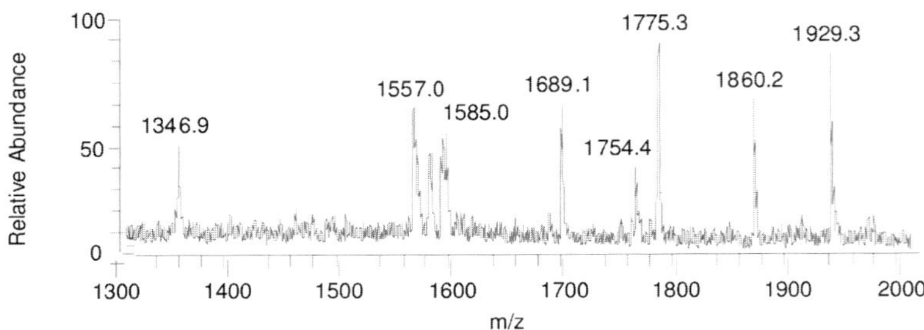

Figure 8. AP-MALDI-MS of aerosolized mixture of BG, MS2 and Ovalbumin.

the collection efficiency and APS measurements, we estimate the spore amount on the spot was around $5 \cdot 10^5$ cfu.

While experiments with pure BG aerosol in water was feasible, such experiments were not applicable to MS2 bacteriophage or ovalbumin. Pure MS2 and pure Ovalbumin likely form particles much smaller than the minimum detectable size by APS (0.5 μm). As shown in Figure 6, particles in this size range are also not collected efficiently. However, if MS2 or Ovalbumin are aerosolized in PBS solution, biological material and the added salt form larger (0.5 μm to 2.5 μm) polydisperse particles which are efficently collected. Once deposited, the wash-step in the sample processing removes most of the salt from the sample. Peptide markers of MS2 and Ovalbumin were observed in such experiments (data not shown).

When a mixture of BG, MS2 and Ovalbumin were deposited, it was possible to detect and identify the individual biological materials. Figure 8 shows the AP-MALDI-MS mass spectrum for collected material on the tape surface processed with the unified broadband protocol as described earlier. Sampling time was 2 minutes. The concentration was around 1000 particles/L (ppL) for the aerosolized material. One can recognize the peptide fragments from BG at *m/z* 1585 and 1929; from MS2 at *m/z* 1755 and from Ovalbumin at *m/z* 1347, 1557, 1689, 1860. The peak at *m/z* 1775 is a trypsin autolysis product.

Differentiation of Subtypes of Organisms

Here we describe using MS/MS typing based framework for detection and differentiation between BA and its close neighbors, which exemplifies the usage of the method. Spectral comparison of MS/MS observed at *m/z* 1518.67 to the MS/MS signature present in the MS/MS reference library leads to the detection of peptide LVSLAEQQLGGFQK in the sample, which in turn results in the detection of a first tier of candidate organisms which contain this peptide. The list of such organisms, according to Rapid Microorganism Identification Database (http://www.rmidb.org) as of May 2009, is as follows:

Bacillus cereus ATCC 10987 (taxonomy: 222523)
Bacillus cereus B4264 (taxonomy: 405532)
Bacillus cereus NVH0597-99 (taxonomy: 451707)
Bacillus cereus W (taxonomy: 405917)

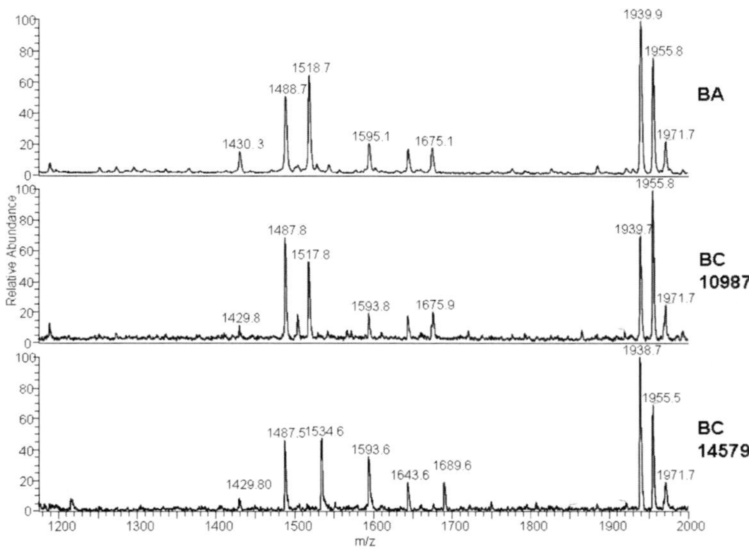

Figure 9. AP-MALDI-MS Spectra of BA, BC10987 and BC14579 extracted under SASP chemical protocol (target specific for spores).

Further analysis as of May 2009, NCBInr search also suggests that the following organisms may share this peptide:

Bacillus cereus Rock3-42 (taxonomy: 526985)
Bacillus thuringiensis serovar pakistani str. T13001
Bacillus thuringiensis serovar pulsiensis BGSC 4CC1

In the second tier of the detection, we must use another biomarker which is not present in all or some of the listed organisms. Thus, another MS/MS data for the parent ion at *m/z* 1527.8 is examined and the detection of peptide LVSLAEQQLGGGVTR eliminates *Bacillus cereus* ATCC 10987 from the list of candidates. The combination of fragmenting *m/z* 1518.67 and *m/z* 1527.8 allows narrowing down the list of candidates to the 6 above shown possibilities.

To further narrow down the identifications, the sequences of the candidate organisms were examined for any other biomarkers absent in these organisms. While there is theoretical sequence difference which would allow for more differentiation, we also inspected measured MS spectra of BA versus the two available BC strains for specific biomarkers of non-peptide nature. This can be particularly useful in cases when there are no peptide based biomarkers available (all the peptide sequences are non-unique to our species of interest). As expected, spectra of BA vs. BC processed under SASP protocol are extremely similar (Figure 9). The difference between BA and BC 14579 in peaks 1518 and 1534 allows for differentiation. The presence of a small peak at 1527 in BA allows for differentiation between BA and both BC strains. However, the majority of other peaks (identifiable or not) are the same and the extreme similarity between SASP peptides in close neighbors leads for need to explore additional strategies. We use

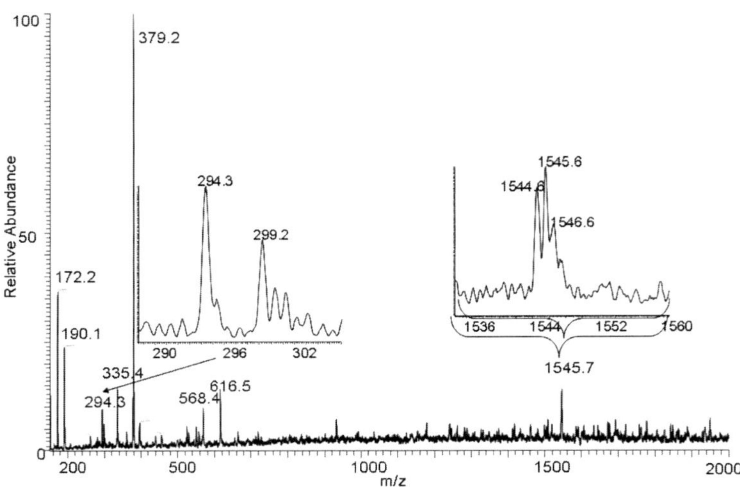

Figure 10. AP-MALDI-MS spectrum of BA spores under whole spore analysis protocol.

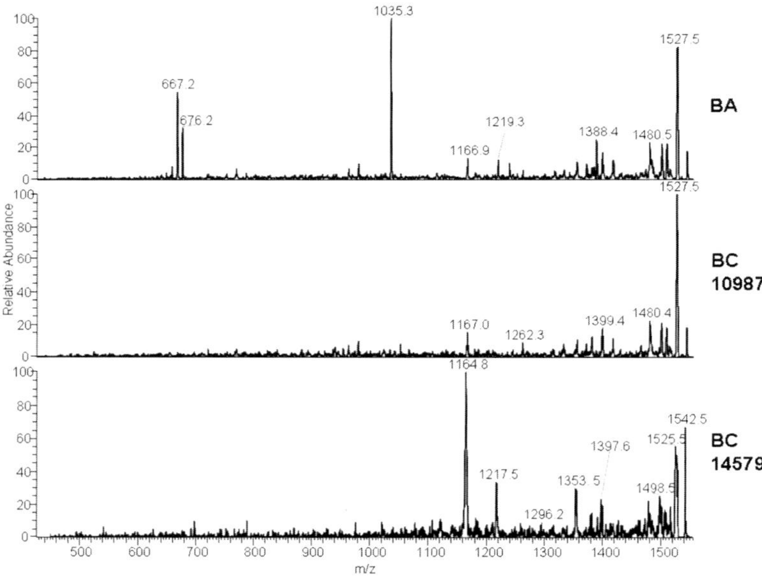

Figure 11. MS/MS spectra of m/z 1545 for spores BA, BC 10987 and BC 14579

an additional chemical protocol to search for new biomarkers not associated with SASP proteins, which, in conjunction with the SASP protein detection (which are used for the first tier of detection) can provide us with unambiguous identification. To solve this, a combination of SASP biomarkers and any experimentally observed species-specific biomarkers is used for unambiguous identification.

A further step was taken to extend the differentiation between the organisms, by applying a different chemical protocol – a whole spore analysis protocol.

The whole spore protocol involves adding matrix to the spores, without any extraction or digestion step and is expected to produce a spectrum with spore surface proteins/peptides. Figure 10 displays the MS spectrum collected under the whole spore protocol and contains peaks which are likely not SASP proteins. Similar analysis for BC spores (close neighbors) was also performed for this protocol. Magnified peak at *m/z* 1545 present in BA spores was not observed in BC strains, as well as the peak at *m/z* 294. The identity of *m/z* 1545 is not known, but under close observation it was also found under SASP extraction protocol (although with very low intensity). MS/MS patterns for m/z 1545 from BA were distinct from those of BC strains. Therefore, we hypothesize that the observed peak at 1545 present in BA spores likely belongs to a non-peptide biomarker which should not be detectable by MS/MS search but can easily be detected by MS/MS typing.

The fragmentation spectra of precursor 1545.6 for all three organisms are shown in Figure 11. Although some similarity is observed, it is obviously different between BA and both BC strains. MS/MS typing can easily distinguish between these patterns. As expected, MASCOT search did not result in any identification, probably due to their non-peptide nature and therefore its absence in the protein sequence database. In detection mode, this type of target specific analysis takes 2 to 3 minutes if sample processing is excluded. It required combination of 2 MS/MS spectra and their analysis which took 2 minutes to collect and a few milliseconds to analyze.

Summary

AP-MALDI-MS based detection approach is shown to analyze aerosolized biological material. Differentiation of close relatives of BA and BC is shown using non-protein markers.

Acknowledgments

The authors thank initial funding for the detection system development by US Army RDECOM. SESI IR&D funds provided further development including miniaturization. The authors thank Prof. Catherine Fenselau and co-workers for helpful contributions and discussions.

References

1. Inglesby, T. V.; Henderson, D. A.; Bartlett, J. G.; Ascher, M. S.; Eitzen, E.; Friedlander, A. M.; Hauer, J.; McDade, J.; Osterholm, M. T.; O'Toole, T.; Parker, G.; Perl, T. M.; Russell, P. K.; Tonat, K. *JAMA, J. Am. Med. Soc.* **1999**, *281*, 1735–1745.
2. Henderson, D. A.; Inglesby, T. V.; Bartlett, J. G.; Ascher, M. S.; Eitzen, E.; Jahrling, P. B.; Hauer, J.; Layton, M.; McDade, J.; Osterholm, M. T.; O'Toole, T.; Parker, G.; Perl, T.; Russell, P. K.; Tonat, K. *JAMA, J. Am. Med. Soc.* **1999**, *281*, 2127–2137.

3. Audi, J.; Belson, M.; Patel, M.; Schier, J.; Osterloh, J. *JAMA, J. Am. Med. Soc.* **2005**, *294*, 2342–2351.
4. Sun, Y.; Ong, K. Y. *Detection Technologies for Chemical Warfare Agents and Toxic Vapors*; CRC Press LLC: Boca Raton, FL, 2005.
5. Demirev, P. A.; Fenselau, C. *J. Mass Spectrom.* **2008**, *43*, 1441–1457.
6. Kientz, C. E. *J. Chromatogr., A* **1998**, *814*, 1–23.
7. Warscheid, B.; Fenselau, C. *Anal. Chem.* **2003**, *75*, 5618–5627.
8. Warscheid, B.; Jackson, K.; Sutton, C.; Fenselau, C. *Anal. Chem.* **2003**, *75*, 5608–5617.
9. Sundaram, A. K.; Oktem, B.; Gudlavalleti, S. K.; Razumovskaya, J; Gamage, C.; Kurnosenko, S.; Serino, R. M.; Doroshenko, V. M. *Anal. Chem.*, submitted.
10. Pribil, P., A.; Patton, E.; Black, G.; Doroshenko, V.; Fenselau, C. *J. Mass Spectrom.* **2005**, *40*, 464–474.
11. Marple, V. A.; Chien, C. M. *Environ. Sci. Technol.* **1980**, *14*, 976–985.
12. Cheng, Y. C.; Yeh, H. C. *Environ. Sci. Technol.* **1979**, *13*, 1392–1396.
13. Pak, S. S.; Liu, B. Y. H.; Rubow, K. L. *Aerosol Sci. Technol.* **1992**, *16*, 141–150.
14. Cohen, S. L.; Chait, B. T. *Anal. Chem.* **1996**, *68*, 31–37.

Chapter 12

MALDI Mass Spectrometry for Rapid Detection and Characterization of Biological Threats

Nathan A. Hagan,[*,1] Jeffrey S. Lin,[1] Miquel D. Antoine,[1] Timothy J. Cornish,[2] Rachel S. Quizon,[1] Bernard F. Collins,[1] Andrew B. Feldman,[1] and Plamen A. Demirev[1]

[1]Johns Hopkins University Applied Physics Laboratory, Laurel, MD 20723
[2]C&E Research, Columbia, MD 21045
[*]nathan.hagan@jhuapl.edu

The Chemical/Biological Time-of-Flight (CB-TOF) is an end-to-end mass spectrometry-based system for rapid triage of "white powder" potential biological and chemical threats. CB-TOF utilizes matrix assisted laser desorption/ionization (MALDI) mass spectrometry (MS) as an analytical tool to detect and characterize intact microorganisms and biotoxins (e.g., *Bacillus anthracis* spores and ricin, respectively) in less than 35 minutes. The system includes a robotic station for rapid, repeatable sample preparation for MALDI MS, a commercial laser desorption time-of-flight mass spectrometer, and novel detection algorithms, combined with a high-level technician-friendly graphical interface. The algorithms for bio-threat detection and presumptive identification are based on both empirical MS signatures and available proteomic database information. A large number of samples of appropriate bio-threat simulants have been examined, allowing characterization of the CB-TOF system performance, e.g., probabilities of detection and false positive rates, and differentiation of spore near-neighbors in a blind study. Confirmation of initial spore presumptive identification has been achieved through growth and vegetative biomarker detection. The transferability of CB-TOF protocols and database signatures to other laboratories has been evaluated as well.

Introduction

At the end of 2008, a bipartisan Commission on the Prevention of Weapons of Mass Destruction (WMD) Proliferation and Terrorism, appointed by the US Congress, released a report titled "World at Risk" (*1*). This report states that: "…it is more likely than not that a WMD will be used in a terrorist attack somewhere in the world by the end of 2013. The Commission further believes that terrorists are more likely to be able to obtain and use a biological weapon than a nuclear weapon…." (*1*). Unfortunately, the production and deployment of biological weapons on a large scale is well within the reach of rogue states, trans-national or regional terrorist groups, and even lone extremists with modest skills in microbiology/biochemistry. In that context, there are typically more than a thousand "white powder" hoax incidents per year in the United States (*2*). Effective and efficient countermeasures against such perceived or real bio-threats require the development of methods for their rapid and robust detection and characterization.

Compared to other technologies for bio-threat detection and characterization, mass spectrometry (MS) presents several unique advantages (*3–5*). In MS, rapid detection and identification of bio-threats – both microorganisms and toxins – is achieved by detecting the masses of unique biomarkers that can be correlated to each threat. MS methods can rapidly provide strong evidence for a microorganism's identity on species level within minutes to hours, which is comparable to time requirements for PCR, and is much faster than the days that are needed for classical microbiology methods. MS is broadband since it can be applied to all classes of bio-threats – from small molecule toxins to viruses to eukaryotic parasites. For example, pure toxins (e.g., ricin or botulinum neurotoxin) do not contain DNA. Thus, all technologies based on PCR and DNA sequencing would have only limited application for toxin detection. In addition, because the sets of PCR, immuno-assay, and MS inhibitors are not identical, orthogonal detection methodologies are needed for maximum detection probability. MS is sensitive – with limits of detection reported for MALDI MS of less than 10^4 intact organisms (*6*), and picomoles to femtomoles of protein toxins, depending on toxin type and the specific assay employed (*7–9*). However, additional efforts are required before MS can be moved out of the analytical/academic laboratory and become a useful tool for non-experts. Results from the development of such an integrated system – the CB-TOF, suitable for deployment in a "front-line" (e.g., a state public health) laboratory – will be reported here.

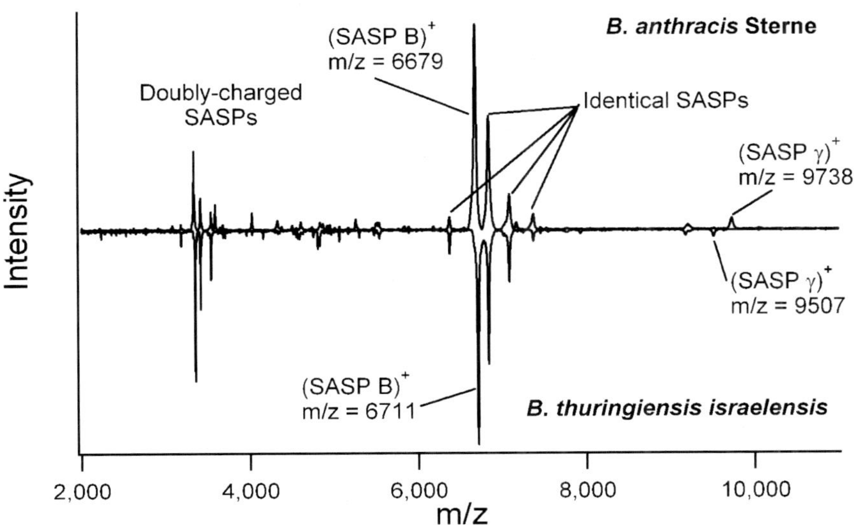

Figure 1. Comparison of positive ion MALDI mass spectra from intact spores - Bacillus anthracis Sterne and Bacillus thuringiensis. Only two of the major observed biomarkers (SASP B and SASP γ) differ in mass between these two species, as predicted from their respective genome sequences.

MALDI MS for Intact Microorganism Detection

In MALDI, an appropriate photo-absorbing organic compound (matrix) is mixed with the sample (e.g., intact bacterial spores) prior to introduction into the mass spectrometer. Sample irradiation with a pulsed ultraviolet or infrared laser desorbs high-mass bio-molecular ions for subsequent MS analysis. Acquisition of threat-specific mass spectral information - a "signature" - is the basis of MS for detection and identification of bio-threats. As with any sensor applied for chem- and bio-defense, signature uniqueness (specificity) is one of its most important features. In general, pathogenic bacteria introduced intact in a mass spectrometer generate unique signatures that allow taxonomic distinctions to be made between different organisms (*3*, *4*). Currently, proteins are the most reliable biomarkers for detection and characterization of both microorganisms and toxins. Therefore, MS-based proteomics combined with bioinformatics is particularly well suited for bio-defense applications.

In Figure 1, the MALDI MS signatures from two intact spore species - *Bacillus anthracis* Sterne and *Bacillus thuringiensis* - are compared. The two species are close "relatives" since a large portion (~ 85%) of their genomes are identical. In this case, the major biomarkers detected by MALDI MS from the intact spores are small acid-soluble spore proteins (SASPs), found in high abundance (up to 10% by weight) in dry spores (*6*, *10–14*). Due to the genomic similarity between the organisms, only two of the major SASPs differ in mass between the two species (*11*, *14*). These differences are preserved across known strains of the two species, allowing unambiguous differentiation and confident identification to be made. Lipopeptides, which are a class of biomarkers that are not predicted from the respective genome, can also be detected in spectra of intact spores at around mass-to-charge, *m/z*, 1500 (not shown). Other classes of highly abundant proteins such as ribosomal proteins have been utilized as biomarkers in MS signatures from vegetative bacterial cells (*15*). MALDI MS for microbiological applications has been commercialized, including appropriate signature libraries and search algorithms (*16*). While these systems are aimed at clinical applications, they can be successfully applied for bio-threat detection. In addition to microorganisms, intact molecular weight information for toxins, e.g., ricin, is also most readily and most rapidly obtained by MALDI MS, allowing sample homogeneity to be assessed.

CB-TOF System Hardware

The CB-TOF system is an "off-spring" of the Bio-TOF system, also developed at the Johns Hopkins University Applied Physics Laboratory (JHU/APL), and is based on MALDI time-of-flight (TOF) MS (*17*). The CB-TOF includes a sample preparation station (SPS) for rapid and repeatable sample preparation for MALDI MS analysis, a commercial laser desorption TOF mass spectrometer, and in-house developed novel detection and signature matching algorithms, combined with a user-friendly graphical interface - Figure 2.

Sample Preparation Station

The SPS (prototypes built by Prototype Productions, Inc., VA) is a custom-built programmable robotic sample pipetting system built around a commercial multi-channel pipette. The main SPS components are: the pipette head, sample cartridge with disposable plastic modules, the pipette position system, and the heating element. The plastic sample cartridge is designed to house all necessary reagents required for MALDI MS analysis, as well as to contain all disposable components that may become contaminated due to contact with the sample. The powder sample suspended in a liquid is introduced by emptying a vial into a front trough. Specially developed and optimized solvents, matrix solution, etc., are all contained in the sample cartridge. Pipette tips, which are used to transfer the sample and reagents between wells and to the MALDI target, are picked up from and returned to positions in the sample cartridge. Sample preparation is performed as previously described (*20*). Briefly, 1μL of α-cyano-4-hydroxycinnamic acid matrix, sample, and 10% trifluoroacetic acid are successively deposited on the MALDI target. Evaporation of the solvents is facilitated by an adjustable heating element. The entire sample preparation procedure is automated and takes around 15 minutes (depending on number of spots and drying conditions). The SPS accepts industry-standard 96-well MALDI targets that can be introduced into a variety of commercial MS systems for subsequent analysis. The SPS is programmed via text scripts to automatically perform a number of biochemical analysis operations and to deposit samples using a variety of sample prep protocols.

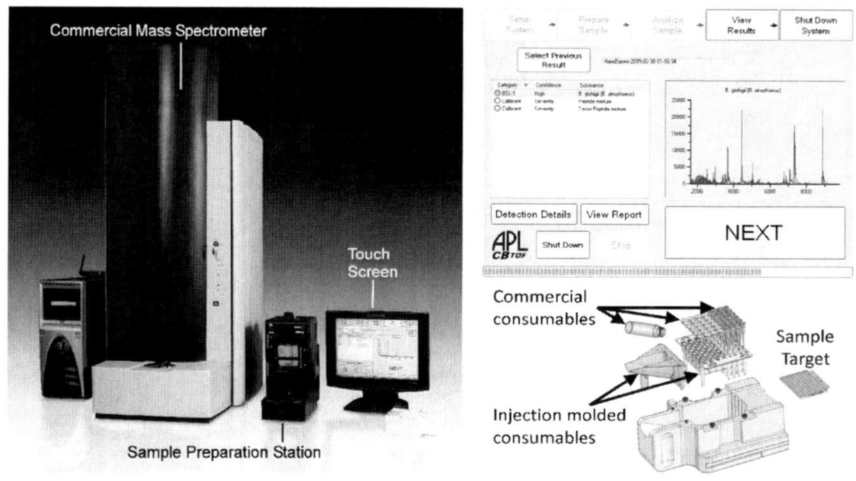

Figure 2. CB-TOF prototype system, sample cartridge components, and a screen-shot illustrating the customized GUI.

Time-of-Flight Mass Spectrometer

Automated MALDI-TOF mass spectral analysis is conducted on a commercial bench-top mass spectrometer, a Bruker Microflex (Bruker Daltonics, MA). The Microflex instrument is equipped with a cartridge-based N2 laser (λ = 337 nm), 4 ns pulse width, ~ 150μJ pulse energy, 20 Hz max repetition rate). Mass spectra are acquired without user interaction in linear positive ion mode utilizing automated routines built into the instrument acquisition software (FlexControl, Bruker Daltonics, MA). Each average mass spectrum consists of the sum of several hundred individual spectra, each acquired from a single laser shot, uniformly covering an entire sample spot. This average spectrum is then automatically smoothed and background subtracted using the software provided by the instrument manufacturer. Extracted biomarker peaks are then automatically analyzed and compared by the CB-TOF software with known biomarkers for the simulant sample. Different low cost MALDI-TOF MS instruments from other commercial vendors can be integrated in a CB-TOF system.

CB-TOF System Algorithms and Software

Signature Library Generation

The algorithms that are used for biological agent detection are comprised of several discrete computational steps. The first is the detection of peaks in individual mass spectra. Here we leverage the peak detection algorithms built into the software of the commercial mass spectrometer. The software provides a list of peaks, where each peak is characterized by its assigned m/z ratio and peak height. We have developed our own algorithm to estimate the signal-to-noise ratio (S/N) for the peaks and to effectively "correct" the height of each peak with increasing m/z. MALDI by its nature is stochastic, with numerous chemical and physical processes interacting non-linearly to generate a complex mixture of ions and neutrals from the solid. As a result, not all peaks from the same sample are evident in every spectrum and the relative peak heights vary over a moderate range. The one invariant, however, is the mass of the constituent peptide and protein biomarkers. Experimentally it can be determined only with a limited accuracy. Thus, an empirical signature must reflect the probability that a peak will be observed, the reliability of its height relative to other peaks, and the variability of the mass assigned to the peak. We have developed a procedure for generating an empirical signature that characterizes each biological agent (or simulant) to account for this particular nature of MALDI mass spectra. The signature is an ordered list of mass intervals ("bands") around each biomarker mass, with a normalized weight ("importance" value) for each interval that reflects both the observed peak amplitude and the relative frequency of peak detection in multiple spectra from the same sample. The signatures are derived empirically from spectra obtained under test conditions similar to those expected during field usage. All generated signatures are stored in a signature library. The peaks observed in a spectrum of an unknown are compared against the library of generated signatures for various toxins and microorganisms. These signatures can be also derived *in silico* by combining an assumed instrument precision with the known molecular weight of a toxin, or from a proteomic analysis of genetically sequenced microorganisms (*18*, *19*). These two approaches can also be combined, with biomarkers found/confirmed by both methods assigned higher statistical weights when comparing to spectra of unknown samples. The protocol allows subsequent library expansion and update to include additional targeted threat signatures.

Detection of Unknowns

For detection of an unknown, we count the number of matched peaks in its spectrum to a given signature's bands in our library. At the certainty level sufficient for biological agent detection, we have developed algorithms to estimate the false-match probability that an observed spectrum matches by chance a signature other than the one for the presumed source (*19*). To quantify it, we introduce the p-value: the probability that a uniform distribution of the same number of peaks observed in the spectrum of the unknown will have at least the observed number of matches between the unknown spectrum and each signature. A lower p-value corresponds to higher certainty of a correct match (detection). We further improved our algorithms in order to use the informative, but not deterministic, relative importance of biomarker peaks. For that, we multiply (logical AND operation) the false-match probability p-value with the Spearman rank-order correlation coefficient to yield the modified p-value of the observed spectrum, π. We then convert π to a score, $s = -\log10(\pi)$. The log transform compresses the dynamic range of the scores and a higher score number signifies a more significant match. The false-match probability is calculated for a set of observed spectral peaks and a set of signature biomarker mass ranges. A key difficulty for calculating this probability is in selecting how many spectral peaks and how many signature mass bands to include. The S/N of the spectral peaks may fluctuate from very high for high concentrations of analyte to very low for trace amounts. Likewise, an empirically derived signature can contain an arbitrary number of mass bands. We have developed an algorithm to calculate the highest probability of detection using all possible combinations to avoid arbitrarily selecting S/N thresholds or absolute counts for either the spectrum or signature. All biological agents/simulants with detection scores above a lower cutoff value are reported to the user.

Identification of Unknowns

Related *Bacillus* species, such as those in the *B. cereus* family, have correlated genomes, and therefore correlated SASP sequences/masses. This is illustrated in Figure 1 for the closely related *B. thuringiensis* and *B. anthracis* Sterne. This non-random partial overlap of signatures leads to correlated detection scores. For this reason, it is possible that several related microorganisms could be automatically detected in a sample containing only a single microorganism. A disambiguation (presumptive ID) algorithm was developed to discriminate between related entries in the library using a "winner-take-all" strategy in assigning spectral peaks obtained from an unknown sample to library signatures. The library signature with the highest original detection score "claims" all matching peaks in the spectrum of the unknown, leaving the remaining peaks to be claimed by other library signatures. Thus, the presumptive identification score is reinforced for the library signature best matching the data, while scores of near-neighbor organisms with overlapping signatures are discounted. Library signatures with sufficient independent support in the spectrum can "compete" for the remaining spectral peaks, allowing the presumptive identification of multiple unrelated organisms in a mixed sample. The presumptive ID scores, which range from 0 to 1, are reported to the user. (Further details of the presumptive ID algorithm beyond the scope of this manuscript will be reported elsewhere.)

Graphical User Interface (GUI)

The CB-TOF operator can control both the sample processing station and the commercial mass spectrometer from a single high-level user interface. The interface was designed after discussions with representatives from various government hazardous materials response teams. The step-by-step "install wizard" level software provides both concise written instructions and short videos describing the tasks required to complete the analysis of an unknown sample. The GUI includes HTML reporting capability and live video/still archiving of sample preparation. Once the sample is fully processed in the SPS, the software instructs the user how to transfer the MALDI sample target from the sample cartridge and place it in the mass spectrometer. After a single button press on the touch screen (or a single click of the mouse), the sample is then automatically analyzed and the previously described algorithms automatically process the data. All results of the algorithmic analysis of the spectra are summarized for the user. The list of results is ranked and color-coded depending on the hazardous/threat level of the presumed detection and presumptive identification. Failures of the system to detect correct responses from the negative or positive controls or to perform calibration are highlighted as well. In addition to the tabular detection summary, a drill-down option of the data/results is available for on-site or for remote inspection. It includes a spectral exploration tool, allowing further and more detailed examination of the spectrum by an expert, a reference spectrum for the detected bio-threat, the peaks used to justify the detection, and the signature bands used to identify the threat. All data, including spectra, detection scores, user-supplied sample information, and sample preparation videos, are archived for future analysis.

Results

Blind Study: *Bacillus* Spore Discrimination

In order to test the effectiveness of the CB-TOF system for rapid triaging of spores, we performed a blind study to evaluate its limits-of-detection (LOD), false-positive/false-negative detection rates, and presumptive identification (ID) rates. Ten *Bacillus* spore species (members of inclusivity as well as exclusivity test panels) were grown in four different growth media (Tryptic Soy Broth, TSB; Nutrient Sporulation Media, NSM; Brain Heart Infusion, BHI; Yeast Extract/Tryptone, 2xYT) according to standard procedures (*6*) to determine the effect of growth media on spore signature variability. The cultures were washed with water to remove excess media, independently enumerated by microscopy, and then presented blind for an end-to-end analysis by the CB-TOF system. The organisms chosen include four close "relatives" belonging to the *B. cereus* group: *B. anthracis* Sterne (Bas), *B. cereus*, *B. thuringiensis israelensis* (Bti), *B. thuringiensis kurstaki* (Btk), as well as six other *Bacillus* species: *B. circulans*, *B. licheniformis* (Bl), *B. megaterium*, *B. mycoides* (Bm), *B. subtilis*, *B. atrophaeus* (*B. globigii*, Bg). Signatures for several species did not exist initially in the CB-TOF library, and signatures of coded "unknowns" were generated from the samples providing the most intense MALDI MS signals.

The results from spore growth in multiple culture media are illustrated in Figure 3. All major SASP biomarkers were detected by MALDI MS regardless of media and no mass shifts of the major biomarker peaks were observed for different growth conditions. Growth media affected growth/sporulation efficiency of *Bacillus*, thus effectively reducing the spore concentration in the sample and the number of spores on the target (for the same deposited sample volume).

When spore count exceeded LOD (previously measured to be ~10^4 spores deposited on the target, as determined by an automated CB-TOF assay detection rate greater than 95%), most samples were successfully detected. 100% of the blind Bas and 94% of the *B. cereus* family samples were correctly detected and presumptively identified both by manual and automated algorithms (Table 1). Two samples in the panel (a Bti and a Bl) were not automatically detected with sufficiently high confidence even though microbiological analysis confirmed that the viable spore count was ~10^5 deposited on the MALDI target. This was due to very poor signal for the SASP biomarkers relative to other confounding peaks in

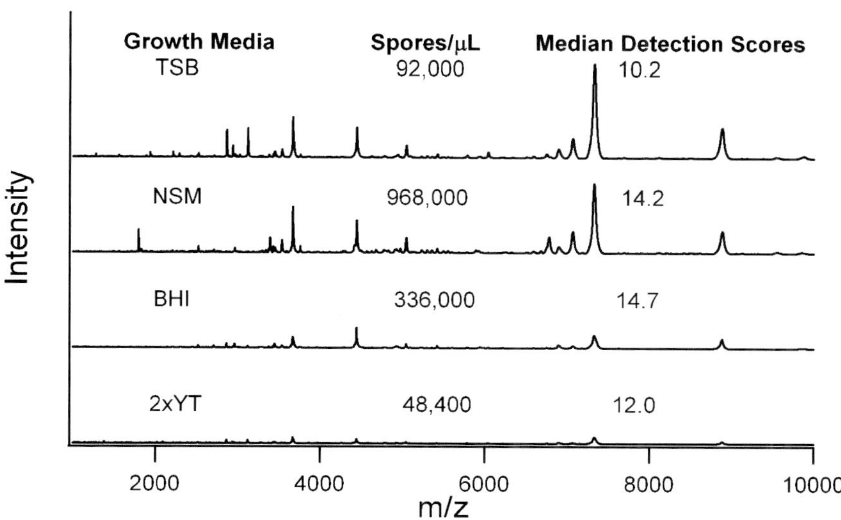

Figure 3. Comparison of MALDI mass spectra of Bacillus atrophaeus spores, grown in four different media.

the spectra. However, these samples could be manually identified by observation of biomarkers in the SASP m/z region and comparison with existing proteomics databases. Conversely, three samples were correctly identified by the automated algorithm as Bm and Bc but could not be manually identified due to lack of proteomic database information. A total of 13 samples could not be detected/ identified either manually or automatically because no SASP biomarkers were observed in the mass spectra. Additional microbiological analysis confirmed that these samples contained a low concentration of viable spores, which resulted in very few spores deposited on the MALDI target (typically $\sim 10^3$ spores or fewer). No mis-calls (i.e., 0% false detection/ID rate) were made by the CB-TOF algorithms, including for non-*cereus* family spores.

Several of the *Bacillus* species/isolates presented to the CB-TOF system during the blind panel analysis had not been previously analyzed on the system. Therefore no signature for those samples had been placed in the library. This provided an interesting case study into how the system may respond to novel or genetically modified organisms that it has never previously "seen". Bl was detected by the system in three out of four cases, but (correctly) could not be identified as any species contained in the library. Rather, the samples were labeled as matching other unknown samples that had been previously analyzed during the blind study (later confirmed to also contain Bl). Btk was also not present in the CB-TOF signature database. However, significant signature overlap with Bti due to nearly identical SASP biomarkers led the presumptive identification algorithm to choose Bti as the signature that best explained the evidence contained in the unknown sample spectrum. While discrimination between Bti and Btk is possible based on a single SASP mass (9507 vs. 9540 Da), this demonstrates the

Table 1. Results of blind study with ten *Bacillus* spore species

Spore	*Automated True Positive Detection/ID*	*Manual True Positive ID*
B. *cereus* family	15/16	16/16
B. anthracis sterne	4/4	4/4
B. cereus	4/4	4/4
B. thuringiensis israelensis	3/4	4/4
B. thuringiensis kurstaki	4/4[b]	4/4
B. circulans[a]	1/4	-
B. licheniformis	3/4[b]	4/4
B. megaterium[a]	2/4	-
B. mycoides[a]	-[b]	-
B. subtilis[a]	-[b]	-
B. atrophaeus	4/4	4/4

[a] Note: No SASP data was found in currently available proteomic databases for several species. [b] Automated calls indicating species that were not originally present in the CB-TOF signature library (see text for details).

robustness of the algorithm to small changes in biomarker peaks that would be expected for genetically similar species/isolates.

Manual Proteomics ID of Unknowns

Mapping (*18*) the spectral peak masses of an unknown sample to a proteomics database of annotated SASP from all sequenced *Bacilli* enabled the tentative identification of the sample as Bl, even though Bl was not in the signature library (Figure 4). In order to quickly confirm this identification without resorting to more traditional bottom-up and/or top-down proteomics approaches, rapid on-target intact spore oxidation with hydrogen peroxide was performed (*20*). This rapid oxidation protocol leads to predominant oxidation of the Met residues found in SASPs. After oxidation, the mass of Met-containing SASPs increases by multiples of 16 Da according to the number of Met residues per protein. Comparing spectra of control and oxidized samples allows one to "count" the number of Met and compare to predictions from proteome databases (*11*). As expected for the unknown tentatively identified as Bl, there was a 100% match between predicted and observed mass shifts (Figure 4, inset).

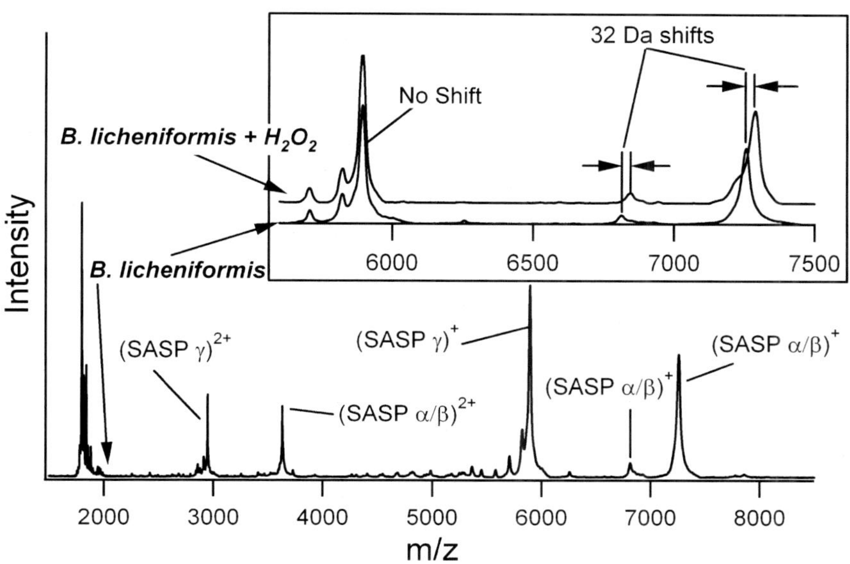

Figure 4. MALDI mass spectrum of a blind unknown sample with peaks mapped to SASP masses from a proteomics database to tentatively identify it as B. licheniformis. Inset: comparison of MALDI mass spectra of the unknown sample before and after rapid on-target oxidation corresponding to the expected mass shifts predicted for B. licheniformis.

Confirmation of *B. anthracis* Spore Detection

To achieve early confirmation of initial spore presumptive identification and determination of viability, we have developed an assay that includes spore germination and cell growth in an appropriate culture medium. Subsequent vegetative *B. anthracis* cell detection in the growth medium is done by the CB-TOF by scoring against the library signature of vegetative cells. As recently confirmed by top-down proteomics and comparison to proteomics databases (*21*), the biomarker signature for vegetative Ba cells contains predominantly low-mass ribosomal proteins. This signature is completely different from the signature for intact spores, and our algorithms allow unambiguous detection and presumptive ID of both spore and vegetative Ba forms in mixed culture samples. Figure 5 shows the CB-TOF detection scores obtained for each time point of a 24-hour culture of Ba spores, along with the corresponding microscopy and cell enumeration. (Note that after 24 hours of growth, nutrients in media were depleted and the organisms began to re-sporulate.) Based on a detection score threshold of ~5, we obtained unequivocal confirmation of the presence of viable Ba cells in the initial sample in less than 5 hours after starting the growth. This time scale is much faster than the typical 12+ hrs required for ID by microscopy/biochemical analysis. The spore growth procedure can be further modified for early detection of *B. anthracis* strain resistance against selected antibiotics.

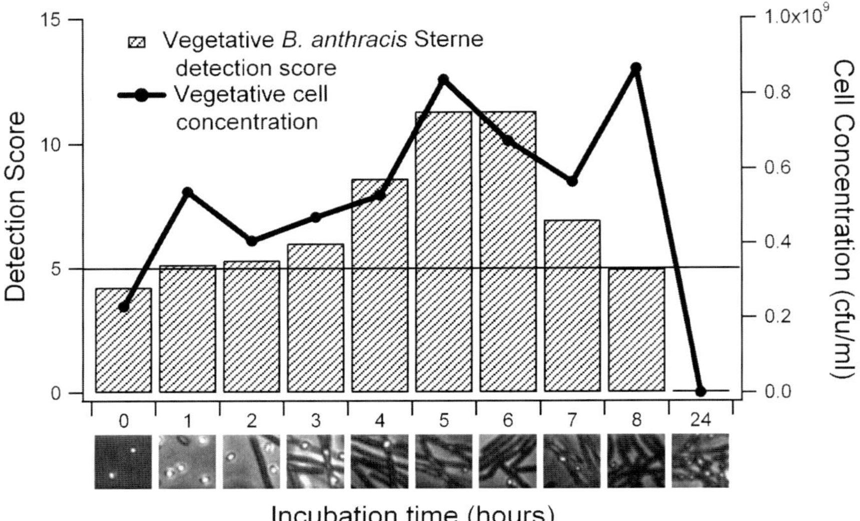

Figure 5. Dependence of vegetative B. anthracis detection scores (log scale axis) and respective vegetative bacterial sample concentration (in colony forming units, cfu, per milliliter) on growth time in nutrient-rich culture.

Conclusions & Future Directions

CB-TOF is a broadband MALDI MS-based system that directly detects proteins from intact cells and toxins and maps them to a signature database. It provides a high-confidence answer in a short time period and at moderate cost. Automated methods enable repeatable and fast end-to-end sample processing, MS analysis, and algorithmic detection. Early triage of suspected bio-terror threats can be performed by CB-TOF in around 30 minutes (vs. ~3 hrs typical for other molecular level sensors). Moderate background tolerance in samples allows rapid triage without sample clean-up. Robust detection and low false positive rates, independent of growth conditions, have been demonstrated. Ultimate sensitivity is not a primary consideration for CB-TOF since a visible amount of "white powder" would contain more than sufficient sample amount for MALDI MS.

Confirmatory chemistries based on MS and proteomics (e.g., functional assays) are available for further integration. MS vegetative cell assays can provide early presumptive identification of a broad range of bio-threats grown in culture, including drug resistance. Independent evaluation of the system in a public health laboratory site has been performed. Ultimately, systems based on MS for detection and presumptive identification of chem- and bio-threats must be incorporated into an integrated threat assessment approach, in which data from orthogonal sensors are fused to provide a high-confidence answer (*22*). As part of an integrated detection suite, CB-TOF can provide such high confidence automated detection and presumptive ID for triage and early confirmation of white powder threats.

Further developments extend the applicability of CB-TOF for triaging of wider range of threats. One such application includes the detection of low volatility chemical threats in powdered form. A problem with the standard MALDI matrixes used for higher mass biomolecule analysis is that they generate a large background in the region of interest for detection of low mass chemicals. Therefore, we have optimized an appropriate non-organic matrix that does not generate background in the MALDI mass spectrum. This matrix allows efficient desorption and analysis of low molecular weight chemicals by the CB-TOF. Pseudo-MS/MS (post-source decay in a reflectron TOF) provides structurally significant fragments that can be used to confirm initial chem-threat detection. Functional and proteomics-based assays for detection of ricin (*7–9*) and other toxins could be implemented on the CB-TOF system. For these as well as for the low mass chem-threat assays, additional work to optimize automated matrix–sample deposition by the SPS, as well as generation of the appropriate library signatures, is required.

Acknowledgments

Funding for this work was provided by the Department of Homeland Security Science & Technology Directorate (A. Hultgren, Program Manager) and by JHU/APL. We also thank J. Becker, J. Hahn, C. Young (JHU/APL), C. Wynne, C. Fenselau (U. of Maryland), J. Barnes, R. Drumgoole, M. Fountain, D. Klein, G. Kubin, L. Su (Texas Department of State Health Services). This work represents the position of the authors and not necessarily that of DHS or any other entity of the US Government.

References

1. *WORLD AT RISK: The Report of the Commission on the Prevention of WMD Proliferation and Terrorism*; Graham, B. (Chairman); Talent, J. (Vice-Chairman); Washington, DC, 2008.
2. Cole, L. A. *The Anthrax Letters*; Skyhorse Publishing: New York, 2009.
3. Demirev, P.; Fenselau, C. *J. Mass Spectrom.* **2008**, *43*, 1441–1457.
4. Demirev, P.; Fenselau, C. *Annu. Rev. Anal. Chem.* **2008**, *1*, 71–94.
5. *Identification of Microorganisms by Mass Spectrometry*; Wilkins, C., Lay, J., Eds.; John Wiley & Sons, Inc.: Hoboken, NJ, 2006.
6. Hathout, Y.; Demirev, P.; Ho, Y.-P.; Bundy, J.; Ryzhov, V.; Sapp, L.; Stutler, J.; Jackman, J.; Fenselau, C. *Appl. Environ. Microbiol.* **1999**, *65*, 4313–4319.
7. Kalb, S.; Smith, T.; Moura, H.; Hill, K.; Lou, J.; Geren, I.; Garcia-Rodriguez, C.; Marks, J.; Smith, L.; Pirkle, J.; Barr, J. *Int. J. Mass Spectrom.* **2008**, *278*, 101–108.
8. Kalb, S.; Barr, J. *Anal. Chem.* **2009**, *81*, 2037–2042.
9. Kull, S.; Pauly, D.; Stormann, B.; Kirchner, S.; Stammler, M.; Dorner, M.; Lasch, P.; Naumann, D.; Dorner, B. *Anal. Chem.* **2010**, *82*, 2916–2924.
10. Demirev, P.; Ramirez, J.; Fenselau, C. *Anal. Chem.* **2001**, *73*, 5725–5731.
11. Demirev, P.; Feldman, A.; Lin, J. *Johns Hopkins APL Tech. Dig.* **2004**, *25*, 27–37.

12. Hathout, Y.; Setlow, B.; Cabrera-Marrinez, R.; Fenselau, C.; Setlow, P. *Appl. Environ. Microbiol.* **2003**, *69*, 1100–1107.
13. Demirev, P.; Feldman, A.; Kowalski, P.; Lin, J. *Anal. Chem.* **2005**, *77*, 5455–7461.
14. Fenselau, C.; Russell, S.; Swatkoski, S.; Edwards, N. *Eur. J. Mass Spectrom.* **2007**, *13*, 35–39.
15. Pineda, F.; Antoine, M.; Demirev, P.; Feldman, A.; Jackman, J.; Longenecker, M.; Lin, J. *Anal. Chem.* **2003**, *75*, 3817–3822.
16. Cherkaoui, A.; Hibbs, J.; Emonet, S.; Tangomo, M.; Girard, M.; Francois, P.; Schrenzel, J. *J. Clin. Microbiol.* **2010**, *48*, 1169–1175.
17. Ecelberger, S.; Cornish, T.; Collins, B.; Lewis, D.; Bryden, W. *Johns Hopkins APL Tech. Dig.* **2004**, *25*, 14–20.
18. Demirev, P.; Ho, Y.; Ryzhov, V.; Fenselau, C. *Anal. Chem.* **1999**, *71*, 2732–2738.
19. Pineda, F.; Lin, J.; Fenselau, C.; Demirev, P. *Anal. Chem.* **2000**, *72*, 3739–3744.
20. Demirev, P. *Rapid Commun. Mass Spectrom.* **2004**, *18*, 2719–2722.
21. Wynne, C.; Fenselau, C.; Demirev, P.; Edwards, N. *Anal. Chem.* **2009**, *81*, 9633–9642.
22. Demirev, P.; Feldman, A.; Lin, J. *Johns Hopkins APL Tech. Dig.* **2005**, *26*, 321.

Editors' Biographies

Plamen Demirev

Plamen Demirev is a Principal Staff member at Johns Hopkins University's Applied Physics Laboratory, which he joined in 2001. He has an M.S. (Physics, University of Sofia) and a Ph.D. (Chemistry, Bulgarian Academy of Sciences). In 1990 he moved to Uppsala University, Sweden, where he became an associate professor. He has performed research at the University of Maryland, UC Berkeley, and Stanford University, and has co-authored more than 100 papers on structural analysis of natural products, ion/solid interactions and mass spectrometry for microorganism characterization. His current interests include physical methods and bioinformatics for rapid detection of chem- and bio-threats in complex environments.

Catherine Fenselau

Catherine Fenselau was one of the first trained mass spectrometrists to join the faculty of an American medical school (Johns Hopkins), where she was given the job description to "exploit mass spectrometry in biomedical research." In early work she applied GC-MS to characterize and quantify achromaphoric metabolites of chemotherapeutic agents, and pioneered MS investigations of intact microorganisms. Subsequent contributions include the novel synthesis and analysis of involatile conjugates of glucuronide and glutathione, measurements of gas phase peptide basicities, proteomic studies of tumor resistance to both chemotherapeutic and immunologic treatment, and rapid characterization of pathogens using mass spectrometry. She is presently a Professor at the University of Maryland.

Indexes

Author Index

Subject Index

C

D

E

F

G

H

I

N

O

P

R

S

T

U

V

Y